The Breakthrough Illusion

THE
BREAKTHROUGH
ILLUSION

■

Corporate America's
Failure to Move from
Innovation to Mass Production

■

Richard Florida and Martin Kenney

BasicBooks
A Division of HarperCollins*Publishers*

Library of Congress Cataloging-in-Publication Data
Florida, Richard L.
  The breakthrough illusion: corporate America's failure to
move from innovation to mass production / Richard Florida
and Martin Kenney.
    p. cm.
  Includes bibliographical references and index.
  ISBN 0-465-00749-X : $19.95
    1. Technological innovation—Economic aspects—United
States. 2. Mass production—United States. 3. Technological
innovation— Economic aspects—Japan. 4. Mass
production—Japan. I. Kenney, Martin. II. Title.
HC110.T4F57 1990   90-80248
338'.064'0973—dc20   CIP

Contents

■

CONTENTS

CONTENTS

CONTENTS

Acknowledgments

■

The idea for this book came to us four years ago when we were both assistant professors at Ohio State University. Since then we have explored many different facets of the sweeping technological and industrial transformation of the late twentieth century: venture capital, high technology in Silicon Valley and Boston's Route 128, corporate R&D, university-industry relations, Japanese industrial organization, and the Japanese transplants in the United States. As much as possible, this book represents a synthesis of those efforts.

We were fortunate to have had the opportunity to discuss our ideas and theories with many academics and nonacademics alike. We owe a special debt of gratitude to the following scholars for the comments and criticisms that helped shape our thinking and make our work stronger: Gordon Clark, Bennett Harrison, Richard Nelson, Shoko Tanaka, and especially Richard Walker, who read the entire draft and provided a wealth of stimulating comments. We would also like to acknowledge a larger network of friends and colleagues for the intellectual stimulus they provided: Harvey Brooks, Frederick Buttel, Wes Cohen, Sam Cole, Phil Cooke, Benjamin Coriat, Marshall Feldman, Robert Gleeson, Norman Glickman, Maryellen Kelley, Linda Lobao, Ann Markusen, Walt Plosila, Ronald Rohrer, Andrew Sayer, Allen Scott, Dale Squires, Michael Storper, Bernadette Tarallo, and Tsuyoshi Tsuru. We greatly appreciate all the help that has been provided by our current and former graduate students, some of whom are now our colleagues: James Curry, W. Richard Goe, Andrew Jonas, and Donald Smith. And we acknowledge the many entrepreneurs, engineers, scientists, corporate executives, and venture capitalists in the United States and in Japan who shared their time and thoughts with us. We must thank our editor, Martin Kessler of Basic Books, for backing this project at an early stage, and Cheryl Friedman, Pat Cabeza, and Susan Woolford for their editing.

The research upon which this book is based has been funded by a number of different sources. Our initial forays were backed by the State of Ohio Urban Affairs Program, the Ohio State University Seed Grant Program, and the state of Ohio. Later funding was provided by the Economic Development Administration of the U.S. Department of Commerce; the National Science Foundation, Division of Geography and Regional Science; the Economic Research Service of the U.S. Department of Agriculture; and the Ford Foundation. Daily support was provided by our various homes, the School of Urban and Public Affairs at Carnegie Mellon University, the Department of Applied Behavioral Science at the University of California, Davis, the Institute of Business Research at Hitotsubashi University, and Ohio State University. We thank them all.

Richard Florida would like to thank his colleagues and staff at the Center for Economic Development for providing a stimulating work environment, and Mary Joyce Airgood for typing and keeping track of administrative details. His deepest thanks go to his wife, Joyce, for her intellectual support and for her love and encouragement, which make all of this worthwhile.

Martin Kenney would like to thank Professors Ken-ichi Imai, Ikujiro Nonaka, and Seiichiro Yonekura for making his stay at the Institute of Business Research of Hitotsubashi University so exciting and useful. He would also like to thank Karen Denney of Ohio State University for typing much of the manuscript.

Finally, we would both like to thank our parents for their assistance over the years.

1

The Breakthrough Illusion

■

In 1986 two IBM scientists working in a Zurich laboratory made an important new discovery. Working with highly advanced ceramic materials, they discovered a new form of electrical superconductor that no longer required extremely low temperatures but could work in real-world conditions. The commercial implications of this revolutionary breakthrough quickly became apparent. The door was now open for major advances in microchip technology, wires, switches, motors, sensors, bearings, lasers, computers, high-powered magnets and motors, even new forms of transportation. The race to commercialize superconductivity was on.[1]

In the United States, IBM led the way. Other large corporations, including AT&T, DuPont, GE, and Westinghouse, quickly followed suit. Seeing the importance of this new breakthrough, the U.S. government began to mobilize funds: the National Science Foundation (NSF) allocated a small amount for basic research, and the Defense Department committed a much larger amount for military applications. In 1987 a bevy of entrepreneurial start-ups with names like American Superconductor Corporation, Conductus, Ceracon, American Magnetics, and Superconductive Components Inc. stepped forward to capitalize on this new technology.[2] Some of these were formed by scientists from corporate R&D labs, but most were established by top university scientists. One of the most important, American Superconductor Corporation, was launched by two MIT scientists, Gregory Yurek and John Vander Sande, and backed by the venture capitalist George McKinney, of Boston's American Research and Development.[3] Only time would tell whether the U.S. could turn its cutting-edge scientific and technological capabilities into a sustained competitive advantage in this important new field.

But in the corporate offices and R&D labs of Tokyo, a different approach took shape. Although they lacked the scientific capabilities of American corporate R&D labs, universities, and spin-off companies, Japanese corporations aimed to capitalize on unique advantages—well-honed capabilities in ceramics and new materials, existing programs in low-temperature superconductivity, and a tremendous capacity to turn technological innovations into products. Electronics giants such as Fujitsu, NEC, Hitachi, NTT, Sumitomo Electric, and Toshiba mounted aggressive product development campaigns. They were joined by companies in other industries: the ceramics firm Kyocera, the steel makers Nippon Steel, Kobe Steel, and Mitsubishi Metals, and Shimizu Corporation, a major builder of high-technology production facilities. These companies recognized the long-term commercial potential of high-temperature superconductivity and invested heavily to develop new superconducting products.[4] The Japanese government helped make this a "collective" effort; in 1988, forty-six Japanese companies joined a government-sponsored consortium, the International Superconductivity Technology Center, a huge research complex in Tokyo. Each company pledged an $800,000 entrance fee and annual dues of $100,000 to develop new superconductor products.[5] By mid-1988 Japanese firms had honed in on a number of product applications: superconducting films, wires, magnets, computer components, and even advanced applications like mag-lev trains, which use magnetic levitation to ride above rails at very high speeds.

As Japan's effort grew, American enthusiasm faded. For large U.S. corporations, the payoffs from superconductor research were slower than anticipated; actual applications were a long way off. Many began to cut back; and those who stuck with it chose to focus their efforts in the laboratory or on specialized military applications. Our fledgling start-ups lacked the production and marketing experience needed to turn their university technology into commercial products.

By late 1988 reports from the National Science Foundation and the Office of Technology Assessment made the new reality painfully

2

clear: in just three short years, the U.S. had fallen far behind Japan in the race to develop superconductor products. The startling findings were reported in a front-page story in the *New York Times* under the headline "U.S. Reported Trailing Japan in the Superconductor Race."[6] By mid-1989 the situation was so bleak that leading U.S. start-ups like American Superconductor had begun to form manufacturing and marketing partnerships with large Japanese companies.[7]

The story of the U.S. superconductor effort is not unique. In semiconductors, computers, and now even biotechnology, much the same thing has occurred.[8] Our strength in basic science, high-end R&D, and other aspects of breakthrough technology has not been enough to hold off foreign, especially Japanese, competitors.

What caused this dramatic turnabout in our technological and industrial fortunes?

The explanation lies in the emergence of a model of technological innovation and development we call the "breakthrough economy." The breakthrough economy has a remarkable capacity to make major new technological breakthroughs, but it neglects the more mundane product and process innovations that are needed to improve new technology, use it effectively, turn it into products, and generate the wealth and economic growth that come from doing so.

A Faltering Position in High Technology

The breakthrough economy is premised on the comforting myth that innovation is synonymous with technological breakthrough—a myth in which most Americans continue to believe. Even though the majority of us are keenly aware of the U.S. decline in automobiles, consumer electronics, and even some older high-tech fields like semiconductors, we take comfort in America's seemingly limitless ability to generate new technological marvels that will keep us well ahead of our major competitors.

In this regard, history seems to be on our side. The legacy of America's ability to develop and, more importantly, commercialize breakthroughs is indeed impressive in the areas of mass-produced automobiles, radio, and television and more recently in high technology. Indeed many of the most important new breakthroughs of the high-technology age have been pioneered by high-flying American enterprises: Fairchild, Intel, and LSI Logic in semiconductors; DEC, Apple, Cray Research, Sun Microsystems, and Compaq in computers; and Genentech, Cetus, Amgen, Integrated Genetics, and Applied Biosystems in biotechnology. Our ability to achieve big breakthroughs is one of the great strengths of our industrial system; in this we are the envy of the world. Lofty academic concepts of product cycles and technology life cycles and the theory of comparative advantage lend intellectual support to the view that we can stay on top by tackling new technological frontiers. A comforting image indeed: the U.S. makes the breakthroughs and forges ahead, leaving older "hand me down" industries to other countries.

But reality is far more complex than these comforting images. Even though many important breakthrough innovations have been made here, the high-technology end products, along with the jobs and wealth they create, are being produced elsewhere—mainly in Japan and the "four tigers" of Korea, Taiwan, Singapore, and Hong Kong.[9] The U.S. continues to lose ground to foreign competitors despite many manifestations of inventiveness and entrepreneurial spirit.

Over the past decade foreign corporations have taken the overwhelming lead in the production of basic semiconductor chips and are gaining in sophisticated custom semiconductors—the much talked about ASIC and RISC chips (application specific integrated circuits and reduced instruction set chips).[10] Japanese corporations now control 40 percent of the world chip market and 70 percent of the market for dynamic memory chips (DRAMs). Fujitsu, NEC, and Hitachi make high-performance supercomputers that are competitive with American-made Crays. Hitachi and Fujitsu produce mainframes that perform as well as IBMs. Portable laptop computers by NEC and Toshiba are among the world's best, and Korean and Taiwanese companies successfully manufacture inexpensive "clones" of our leading personal computers.[11] While U.S. companies continue to lead in the manufacture of hard disk drives, foreign companies already produce 96 percent of the floppy disk drives that go into computers, and Japanese companies continue to build new, highly automated plants to manufacture more floppy and hard drives.[12] In telecommunications, Japanese, and to a lesser extent European, companies are leading manufacturers of fiber optics, private branch exchanges (PBXs), mobile radio systems, local area networks, and fiber optic cable.[13] While the U.S. retains an overwhelming lead in software technology, Fujitsu, Hitachi, NEC, and Toshiba have pioneered automated "software factories" to mass-produce bug-free software programs.[14] And although our start-ups remain ahead in biotechnology, large Japanese and European chemical and pharmaceutical companies are making significant inroads here, too.[15] Worst of all, the U.S. is falling far behind in invention, production, and adoption of important new industrial process technology, including semiconductor production equipment, flexible manufacturing systems, and industrial robotics, which promises to transform manufacturing and create new value, wealth, and employment in the coming decades. Japan is now far ahead of the U.S. in the development of x-ray lithography equipment, which is needed to make next-generation semiconductor chips and advanced computers. According to an NTT researcher, "[The] . . . gap is very, very big. I'm afraid that it is getting bigger."[16]

The numbers speak for themselves. Between 1980 and 1988 (the last year for which data are available) the U.S. high-technology product trade deficit with Japan soared from $3.8 billion to $22.1 billion, while the high-tech deficit with the "four tigers" skyrocketed from $200 million to $10.2 billion.* (See figure 1.1.)

American corporations have only made matters worse by selling more and more advanced technology to foreign competitors. In just the three-year period 1986–88 U.S. companies sold roughly $5.6 billion of technology to Japanese corporations.[17] In fact, revenues from foreign royalties and license fees for American breakthroughs

---

*These data are based on the Department of Commerce DOC-3 definition for high-technology industries, as reported in Department of Commerce, International Trade Administration, Office of Trade and Investment Analysis, *U.S. Trade Performance in 1988* (November 1989). This definition includes the following industries: missiles and spacecraft, communications and electronics, aircraft, office computers, ordinance (military products), drugs and medicine, industrial inorganic chemicals, professional and scientific instruments, engines and turbines, and plastics and synthetics. Since this definition combines the defense and aerospace industries, where the U.S. is a major exporter, with other commercial high-technology sectors, where imports are concentrated, it may seriously underestimate the U.S. deficit in the latter category.

Figure 1.1   U.S. High-Tech Trade Deficit with Japan

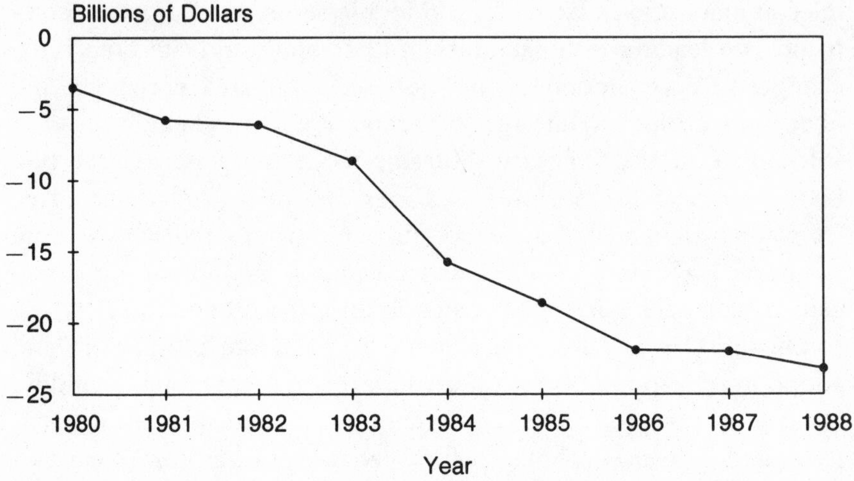

SOURCE: U.S. Department of Commerce, International Trade Administration, Office of Trade and Investment Analysis, *U.S. Trade Performance in 1988* (Washington, D.C., November 1989).

increased steadily during the 1980s, climbing from $12.2 billion in 1980 to almost $40 billion by 1989 (see figure 1.2). Over the decade as a whole U.S. corporations sold more than $225 billion worth of their technology to foreign competitors.[18] The United States now faces a perplexing situation: American corporations sell their breakthroughs abroad, and American citizens buy the foreign products based on that technology.

Our preoccupation with breakthroughs has led us to underestimate the technological strength and innovative capacity of our major rival, Japan. In 1988 the three companies that received the largest number of U.S. patents were Canon, Toshiba, and Hitachi.[19] In 1989 Japanese companies received more than twenty thousand new U.S. patents, more than 20 percent of all patents issued by the U.S. government (see figure 1.3).[20] According to a recent Ministry of International Trade and Industry (MITI) survey, Japanese industrialists thought their country was equal to or ahead of the U.S. in all but one of forty key high-technology areas.[21] The findings of a survey of American executives and R&D scientists are even more ominous. This survey asked two questions: Who is the world's current technology leader? and Who will lead at the turn of the century? About half the

Figure 1.2   Receipts from Royalties and Fees from U.S. Sale of Technology

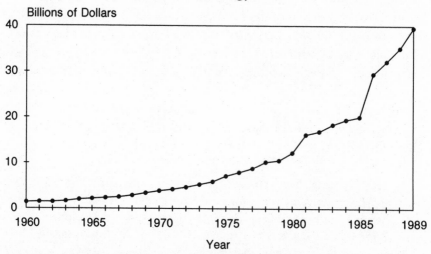

SOURCE: U.S. Department of Commerce, Bureau of Economic Analysis, *Survey of Current Business* (Washington, D.C., various years).
NOTE: Includes royalties, license fees, and other private sources. Data for 1989 are estimates based on receipts through the first three quarters.

Figure 1.3   Growth in Foreign Patents

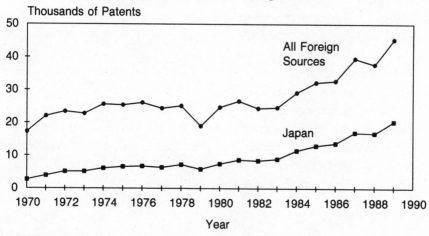

SOURCE: U.S. Department of Commerce, Patent and Trademark Office, *Technology Assessment and Forecast Data Base* (Washington, D.C., January 1990).
NOTE: 1989 data are for fiscal year.

American scientists and two-thirds of the top executives said the U.S. currently leads. But both groups foresaw a gradual erosion of that lead. Only 38 percent of the scientists and one-half of the chief executives thought the U.S. would still be ahead of Japan by the year 2000. In a more recent survey, 69 percent of CEOs of the top 750 U.S. electronics companies thought that the U.S. was losing its technological edge to Japan.[22]

Peter Brooke, founder of TA Associates, one of this country's largest venture capital funds, and financier of dozens of our leading high-technology start-ups, has the following to say on the subject:

> The Japanese have always been good innovators. Although they've not invented as many things as we have, they're inventing more. Their system of applying science is very highly developed— it's true in the markets we compete in. Structurally, institutionally, and socially, that's how they're organized. The U.S. mentality is that it can only happen here—this is the only place with intelligence and cutting-edge technology. That is very dangerous thinking. Our strength is individual invention. Our system doesn't lead to integration and domination in technology development.[23]

The basis of Japan's competitive advantage lies not in breakthrough technology but in its remarkable capacity to make important follow-through innovations in products and manufacturing processes. In this way Japanese corporations quickly turn laboratory technologies into state-of-the-art products and create new high-tech consumer products, office equipment, and industrial machines, thereby outflanking U.S. companies with more advanced next-generation products.[24]

Although the commonplace impression that breakthrough innovations create permanent advantage for American companies may once have been true, it is just not the case anymore. A new reality is upon us: the U.S. makes the breakthroughs, while other countries, especially Japan, provide the follow-through.

What Went Wrong?
Competing Views of Technological Decline

Many have attempted to explain the reasons for America's alarming decline in high technology. One popular explanation suggests that we "gave it away" by selling off technology, promoting free trade even though it hurt key industries, and failing to respond to foreign, mainly Japanese, efforts to subsidize, protect, and enhance key high-tech industries. A leading advocate of this view is a former Commerce Department official, Clyde Prestowitz.[25] Of course, the sale of technology and short-sighted technology and trade policies have contributed to our problems. But the real cause lies in the inability of U.S. high-technology firms to capitalize upon their innovations and turn them into products. In fact, this is largely the reason these corporations have sold off their technology in the first place.

For others, an "excess of entrepreneurialism" is to blame. Robert Reich and Charles Ferguson, two students of the subject, believe that our small entrepreneurial firms cannot survive against Japan's giant integrated corporations.[26] The solution they prescribe is to increase the size of the U.S. high-technology effort through the use of government money and the creation of bigger high-technology companies.

But there is more to it than size.[27] Obviously, small companies find it difficult to compete with Japanese corporations; but big American firms—GE, RCA, Westinghouse, Philco, and Union Carbide—have fared even worse. Even the much-heralded IBM is feeling the heat of competition from both Japanese corporations and American start-up companies. In early 1990, IBM stock plummeted to an all-time low relative to the Standard & Poor 500. "Big Blue" then announced a $2.4 billion "restructuring charge," laying off 10,000 employees and slashing 1989 net income by 35 percent, or $3.8 billion, the biggest drop in the company's history.[28] Simply put: "organization" is a far more important source of competitive advantage than is size. The organization of R&D, the organization of manufacturing, and the integration of the two have powerful effects on the capacity to develop and improve new technology. Big American firms, small high-tech start-ups, and large Japanese corporations are all organized differently, and this—not size—is the crucial factor in their differential abilities to adapt to new circumstances, mobilize resources effectively, generate new innovations, and turn them into products.

Still others explain America's technological decline as the product of deteriorating manufacturing and product development capabilities; that is, U.S. corporations have difficulty producing high-quality, mass-produced goods.[29] Manufacturing certainly matters, and so does product development. But the real problem lies in the failure of U.S. firms to overcome their long legacy of "functional specialization" by developing new functionally integrated ways of organizing R&D, shop-floor production, and connecting them together to tap the full capabilities of all workers. Even the most advanced manufacturing plants will not amount to much if shopfloor workers are not allowed to contribute their knowledge or if such plants are not connected to the innovations that come from R&D labs. Cutting-edge R&D and product development will mean little if they are not connected to manufacturing facilities that can turn innovations into products. The lack of organizational integration in the R&D lab, on the factory floor, and between the two is perhaps the key element in our technological decline, especially since Japanese corporations have pioneered new modes of integration that enable them to generate a continuous flow of innovative new products.

Understanding the Breakthrough Bias

The crux of the situation is that the U.S. technology system is heavily biased in the direction of major new technological discoveries, or breakthroughs. And it is less able to actualize the important follow-through innovations in products and manufacturing processes that are needed to turn new technologies into a constant stream of commercial products.

To comprehend the dimensions of the breakthrough economy, we must first realize that innovation is not synonymous with scientific and technological breakthrough.[30] There are in fact many different types of innovation, all of which are important to a healthy integrated economy. Second, we also must realize that different kinds of innovation require different types of organizations and organizational features. "Incremental" innovations of existing products that enhance performance, reliability, and quality are crucial to making better, more advanced products. The ability to develop better manufacturing techniques or "process" innovations generates much-needed production efficiencies and is a critical contributor to industrial might. Both types of innovation come mainly from manufacturing, not from R&D, but from an alert, educated, skilled force of production workers who can collectively contribute their knowledge and experiment with new ways of doing things. And both depend on tight linkages between manufacturing and product development activities.

Other innovations result from combining technologies into new hybrid products such as fiber optics: the combination of laser and materials technologies.[31] Still others require the integration of a host of separate innovations into a workable system, such as high-definition television, which is based upon multiple innovations—new semiconductor chips, video screens, television receivers, amplifiers, disk storage devices, video cameras, transmitters, and others.[32] Both hybrid and systems innovations depend upon the breadth as well as the quality of R&D and upon close linkages between it and manufacturing. Both require highly integrated organizations that can mobilize financial and human resources to solve complex, multifaceted problems.

And of course, the ability to use advanced technology to create high-tech consumer products, advanced office machines, and new industrial equipment is a form of innovation itself, one that creates mass markets for new technology and provides the profits needed to finance future rounds of innovation. This type of innovation depends on a broad, well-funded R&D effort, integrated manufacturing capabilities, and, more importantly, synergistic interaction between the two.

Because our high-technology system has developed organizational structures that are mainly oriented toward breakthrough innovation, it lacks the capacity to generate the types of innovation that are necessary to capitalize on those breakthroughs. This is a critical failure in an era of rapid, continuous innovation in which what matters most is the ability to produce the latest and best version of a product as

11

quickly and efficiently as possible. Even the most earth-shattering breakthroughs are economically irrelevant if they do not lead to improved products and manufacturing processes or if they fail to transform existing industries.

The key to understanding our current dilemma lies in the way our corporations have chosen to organize R&D and shop-floor production; their failure to move toward functional integration; and their inability to effectively link the sites of innovation and production. These organizational failures take many forms: corporate R&D labs that are located far from factories, the social and physical distance between Silicon Valley and Route 128 and traditional manufacturing industries, and the lack of communication between high-paid white-collar think-workers who work in the gleaming office parks and R&D campuses and low-paid factory workers in the U.S. and increasingly in the Third World. And it is rooted in a long legacy of American organizational practices that separate "brains" and "hands"—managers and R&D scientists, on the one hand, and factory workers, on the other. The lack of integration across the R&D-manufacturing spectrum is the main reason the U.S. is increasingly unable to follow through on the innovations it makes. And it may eventually cost us much of the ability to invent as well.

This book explores the history, organization, problems, and alternatives to the current breakthrough system of U.S. high technology. We begin, in part I, by examining its origins and institutions. Chapter 2 chronicles the problems of the large corporation, the organization of the R&D laboratory, the organization of shop-floor production, and especially the growing estrangement of R&D from manufacturing, which made the new breakthrough model of high-technology organization both possible and necessary. Chapter 3 examines one of the most important elements of the breakthrough model, the high-technology start-up, and shows how it is organized to produce new breakthrough technology. Chapter 4 explores another new institution, venture capital, the finance capital for the breakthrough economy.

Part II turns to the limits and contradictions of this new system. Chapter 5 investigates the hypermobility of high-technology scientists, engineers, and managers, which makes it hard to develop continuity in our high-technology efforts. Chapter 6 probes the weaknesses of the small-firm model of high technology found in Silicon Valley and Route 128, particularly its inability to generate follow-through innovations. Chapter 7 explores the problems that stem from the disregard for manufacturing and manufacturing workers and from the glaring separation of innovation and production in high-technology industry.

Part III highlights the challenges facing the U.S. technology system. Chapter 8 examines the underlying reasons for Japan's remarkable success in high technology—the development of new highly integrative modes of R&D organization, production organization, close connections between R&D and manufacturing, and the ability to follow through on innovation. Chapter 9 surveys the revitalization strategies of both American industry and government and shows why they are not working. Chapter 10 concludes by outlining the deep organizational and institutional reforms that will be necessary to create a stronger, integrated U.S. technology system.

I

▪

ORIGINS AND INSTITUTIONS

2

How We Lost the Follow-through

■

In these large industrial laboratories, . . . research has itself become a mass-production industry. Just as the division of labor in manufacturing made possible significant increases in labor productivity, so the application of this principle to research has improved the effectiveness of the scientific work. . . . [T]he work in these laboratories is systematically divided among specialists in the several sciences or their branches. Each works on a separate phase of the investigation, and few need to know all the phases of the problem.

—1940 government report on the state of industrial research[1]

America was once the world's premier follow-through economy. Its incredible strength was predicated upon three elements: the mass-production assembly line, the industrial R&D laboratory, and an effective method of transferring the results of R&D to manufacturing. The R&D lab was especially important because it gave large integrated corporations a new internal source of innovation. The rise of the modern industrial R&D laboratory enabled large companies to integrate innovation with their already first-class manufacturing capabilities, propelling the U.S. economy to a position of world leadership in previous high-technology industries.[2]

This system had a powerful capacity to turn innovations into manufactured products. Large corporations such as GE, GM, U.S. Steel, DuPont, and AT&T built the world's largest and most modern industrial plants. A new system of assembly-line manufacturing, or Fordism, combined Frederick Winslow Taylor's ideas of scientific management with the orderly flow of the mechanized transfer line.[3] Jobs were broken down into their most elemental components, and the pace of work was dictated by the pace of the line. Productivity increased as more sophisticated and expensive machines were developed and as products and production became more standard and routine. A strict division between shop-floor workers, on one side, and managers and engineers, on the other, was created, with the former workers relegated to the position of hired hands paid to work not to think.[4] A growing managerial bureaucracy was called upon to manage shop-floor labor, coordinate internal corporate transactions, keep things running smoothly, and plan for the future. The vertically integrated corporation managed its various interests as semiautonomous divisions that competed with one another on the basis of standard financial criteria.[5] And a whole new system of investment banks grew up to provide finance capital and promote further consolidation and integration of this new industrial structure.

But gradually, imperceptibly, the United States lost its once powerful follow-through capability. Somehow the same economic structures and institutions that once formed the heart of this advantage could no longer deliver. There were various reasons for this. R&D laboratories became engulfed in bureaucracy and red tape. The desire to maximize short-term profits clashed with the long-term payoffs of R&D. R&D grew further and further estranged from factory production, as companies moved their manufacturing plants to new low-wage, nonunionized locations and then relocated their R&D facilities to suburban campuses. As the gulf between the sites of innovation and production widened, it became increasingly difficult to transform the new technological marvels developed in R&D labs into viable commercial products.

The Industrial R&D Laboratory

The key to the follow-through economy's ability to innovate was the modern industrial research laboratory.[6] The R&D lab fit in nicely with the new model of corporate organization, reinforcing the ideas of vertical integration, functional specialization, and the separation of managers, R&D scientists, and shop-floor workers. The R&D lab was designed to standardize scientific and technical knowledge so that it could be used to improve old products as well as to create new ones. Writing in the 1930s and 1940s, Joseph Schumpeter observed that the corporate R&D laboratory gave American corporations an unparalleled ability to internalize the dynamic of technological change and to turn innovations into downstream products.[7]

The basic model for corporate R&D was Thomas Edison's Menlo Park, New Jersey, research laboratory.[8] For Edison, the R&D lab was a way to make innovation systematic and predictable. He referred to his lab as an "invention factory," announcing his intention to produce "a minor invention every ten days and a big thing every six months or so."[9] Edison's Menlo Park laboratory opened its doors in 1876; by 1892 Edison's General Electric Company was merged with Thomson-Houston, a company founded by the famous inventor Elihu Thomson.[10] In 1900 GE combined its various research activities in a central corporate R&D laboratory, its General Research Laboratory; and in 1911 it established a General Engineering Laboratory to work on product development. Soon a host of other corporations opened their own R&D labs. DuPont opened its Eastern Laboratory in 1902 and in 1903 added a second laboratory, the Experimental Station, to use science to develop new explosives and chemicals as well as to screen outside inventions. AT&T opened a laboratory, the forerunner of Bell Labs, in 1911; and Kodak established one the following year. Westinghouse founded its Forest Hills Laboratory in 1916, while General Motors absorbed Charles Kettering's Delco laboratories in 1920. On the eve of the Great Depression, 115 of the 200 largest companies had established research laboratories, and as many as 1,000 companies had some type of research facility.[11]

The Functions of Industrial R&D

The main mission of most R&D laboratories was to employ science and technology to improve existing products and manufacturing processes. It was gradually recognized that a scientific understanding of the principles and properties of the natural world could generate new commercial products as well. In the words of the famous inventor Elihu Thomson: "A company as large as the General Electric Company should not fail to continue investigating and developing in new fields: there should, in fact, be a research laboratory for commercial applications of new principles, and even for the discovery of new principles."[12] This perspective was shared by a growing number of industrialists, for example, Pierre S. du Pont: "[W]e should at all times endeavor to have in force some investigations in which the reward of success would be very great, but which may have a correspondingly great cost of development, calling for an extended research of possibly several years, and the employment of a considerable force."[13]

Industrial R&D also aimed to enhance the ability of corporations to imitate and learn from their competitors, making American companies quite adept at imitating and improving upon European innovations.[14] In fact, technology-gathering missions to Europe were a commonplace during the late nineteenth and early twentieth centuries, as American corporations sent their leading scientists and engineers abroad to learn advanced techniques. The uncanny ability of American corporations to improve on outside innovations astonished European industrialists in a way that is strikingly similar to the way Japan's success perplexes American managers today. One late-nineteenth-century British writer summed it up: "It is not a very complimentary reflection for European electricians and capitalists that although all ideas and experimental work needed have come from Europe . . . it should be reserved for an American firm to take up the system [of transmitting electric power] and make it the commercial and practical success which the Westinghouse Company is now doing."[15]

And industrial R&D allowed corporations to defend their market positions by creating technological and legal barriers to entry. Patents could be used to create a thicket of barriers to new entrants, temporarily locking out potential competitors. GE, for example, had a stranglehold on the U.S. light bulb market for many years because of its strong patent position. In this sense, corporate R&D had "defensive" as well as "offensive" uses.[16]

During this early, formative period, industrial R&D was quite closely linked to manufacturing and other corporate activities. The great majority of the early R&D laboratories were located either inside actual factories or on factory grounds, creating an intimate interplay between the sites of production and innovation that made U.S. corporations among the world's most competitive and helped them overtake their British competitors, who for the most part failed to achieve this sort of synthesis.[17] There was, furthermore, a correspondence of interests and outlook between the "practical" inventors of the R&D lab and the "practical" managers of the corporation.

How We Lost the Follow-through

Success as a follow-through economy was relatively short-lived, however. By the 1960s signs of stress were visible, and by the 1970s this once-powerful system began to unravel. Initially, only older industries were hurt. Later these problems would affect new high-technology industries as well.

The increasingly specialized and hierarchical division in the R&D lab was the first, and in many respects the most fundamental, problem that emerged. For a time the "founding fathers" of corporate R&D, men like Mervin Kelly of Bell Labs, avoided the extreme functional specialization of Taylor's approach and created a more cooperative environment in their labs. But the R&D lab was gradually subjected to the same techniques of "scientific management" used in other parts of the corporation. There were two dimensions to this: (1) R&D labs were divided into various functional specialties, and (2) they were strictly segregated from later-stage, scaleup, and production activities.

Functional specialization was the cornerstone of postwar R&D organization. Corporations sought to make R&D as standardized, specialized, and predictable as they could.[18] R&D was divided by discipline or specialty, and scientists were organized in groups of specialists with specific roles on well-defined projects. Professional managers increasingly assumed control of research operations, and communication within the lab increasingly was routed through formal bureaucratic channels. Within the laboratory, projects would be passed in assembly-line fashion from researcher to researcher and group to group. Researchers were neither expected nor in many cases even allowed to overstep the boundaries of their particular specialty. Project information was supposed to be passed up the ladder only, not laterally to other parts of the laboratory. For example, in the mid-1950s RCA reorganized its laboratory system to implement this model. Thus, each lab specialized in one area or scientific discipline, for example, physics, chemistry, electronics, or acoustics. Researchers were given the choice of redirecting their work or relocating.[19] Hiring practices also changed to reflect the growing penchant for specialization. At Westinghouse, for example, lab managers could no longer hire generalists; they were required to hire narrow specialists.[20] The R&D lab gradually assumed a level of functional specialization that mirrored that of the factory floor. David Noble summarizes this new organizational system: "As industrial research laboratories grew in size, the role of scientists within them came more and more to resemble that of the workmen on the production line and science became essentially a management problem . . . the industrial researcher was more commonly a soldier under management command, participating with others in a collective attack on scientific truth."[21]

R&D was also separated from other corporate activities such as product development and manufacturing. In 1925 Bell Labs was organizationally separated from Western Electric "to permit more effective specialization in research and development."[22] R&D now became the first step in a specialized assembly-line process of innovation. According to the historian George Wise: "At subsequent workstations along that assembly line, operations labelled applied research, invention, development, engineering, and marketing transform that [scientific] idea into an innovation."[23]

As this process moved along, projects and products would simply be passed over the transom from R&D to product development, from product development to pilot production, and from pilot production to manufacturing. Once a project was handed on, the receiving group was confronted with a fait accompli, their freedom of operation constrained by earlier decisions. Each group optimized according to its own situation and not on the basis of the entire product. Delays and redesign at each stage were common. It typically took a very long time to complete this process, and the complexity added by each stage often made the end products very difficult to manufacture. For example, engineers working on the body of a car might design it in such a way as to make proper placement of the engine and steering difficult. The engineers assigned to steering and motor development would then change the design based on their needs. By the end of this process the car would be expensive and difficult to manufacture. Typically, this method yielded results that were expensive and frequently of low quality, for example, the Ford Pinto and the Chevrolet Vega, cars that were designed as lemons.

The rise of the assembly-line model of R&D was not altogether surprising, since the image of the assembly line must have been incredibly powerful in, say, 1950 or 1955. That model of organization had won the war and could pump out consumer products by the millions. Why not extend it to R&D?

Still its costs were substantial. The R&D lab was not and never could be a mass-production operation. To try to make it one was an unfortunate mistake. In his classic book on postwar American capitalism, *Capitalism, Socialism and Democracy,* the economist Joseph Schumpeter commented that "innovation itself is being reduced to routine. Technological progress is increasingly becoming the business of teams of trained specialists who turn out what is required and make it work in predictable ways. . . . Thus, economic progress tends to become depersonalized and automatized. Bureau and committee work tends to replace individual action."[24] Schumpeter went on to say that the bureaucratization of R&D was a large part of the reason that capitalism would eventually give way to more socialistic modes of economic organization.

All of this had extremely debilitating effects on R&D scientists, the inventive talent of the follow-through economy. It was increasingly difficult for innovation and creativity to take place in the environment of the "organization man." Growing numbers of R&D scientists lost their incentive to innovate. They became disgruntled and ultimately burned-out. Donald Miller, a management expert, described their malaise: "Most professionals . . . in R&D, worry about burnout and obsolescence, about their work being increasingly standardized and formatted, about continuing personal development and career growth, about their psychic income. . . . When I visit research and development organizations, I am often overwhelmed by the feeling that I am seeing powerful people who, for some reason, feel powerless."[25] Yet Miller was unable to diagnose the disease he observed. Like so many others, he prescribed more and better management, when what was really needed was a fundamental restructuring of the entire system of corporate R&D.

It was disastrous to disconnect separate R&D from downstream activities. Under the assembly-line model, very little interaction took place or was even desired between R&D and later downstream functions. But to be successful, R&D requires constant communication and cross-fertilization with manufacturing and other corporate functions. Without such interplay, technological marvels remain on the shelf, never to be turned into actual products.

The Bureaucratic Tangle

The rise of a labyrinthine bureaucracy to oversee corporate R&D was an important contributor to the loss of follow-through. Beginning in the immediate postwar years it became exceedingly difficult to maneuver projects through this maze of sign-offs and approvals.[26] R&D became a focal point for power plays, information hoarding, and short-term political gain. Caught up in a Byzantine maze of corporate bureaucracy and individual turf protection, projects could go through cycles of being designated as high priority, put on hold, and then abandoned, or abandoned, revived as high priority, and then put on hold. R&D projects were constantly being lost in the shuffle of gigantic corporate bureaucracies.

Corporations responded to these problems in ways that only made them worse. New levels of managers were added, further expanding the R&D bureaucracy. Many of these new managers came straight from business school and had little expertise in or "feel" for technology. High rates of executive turnover and job-hopping made things even worse. Managers who were constantly on the move had little concern for long-term R&D projects that would incur costs now but yield returns only after they had moved on.

Corporations tried to use standard financial accounting criteria to evaluate their R&D efforts and compare them with other corporate activities, but without success. The exploratory nature of R&D made it very difficult to estimate, even guesstimate, its potential payoffs, much less compare those payoffs with other options. Jack Goldman, former vice president for research at Xerox, explains: "For the first time, top management requires quantification of its investment in research. This is precisely what most of us who are committed to industrial basic research find most difficult to do."[27]

As planning tools, these techniques were simply inappropriate for dealing with the idiosyncratic process of developing new products and processes. American managers did not proceed down this road without warning. Already in the late 1950s and 1960s high-ranking R&D scientists in some of our leading corporations had begun to speak out. Consider the words of Emmanuel Piore, former vice president for research at IBM:

> We have reached a stage in our economic development where we put tremendous emphasis on operations research and upon market analysis—on the whole formal procedure, in a word, of assuring management that the gamble of introducing a new product will be at a minimum risk. But I have yet to see one case in which there has been a profound growth of a company by means of the procedures that are being taught in the nation's business schools. . . . It seems to me that when real growth occurs it is because some person in a laboratory, some technical person usually, senses what the developing needs of the economy are, and some man in management of the concern has the courage to probe the market. Only then does a product begin to sell.[28]

By applying standard financial accounting criteria to R&D, corporations in effect replaced long-range vision with the safe bet. They placed the responsibility for planning their technological future in a set of accounting and financial management techniques, the short-term orientation of which was completely ill suited to the long-term nature of R&D. In their powerful critique of postwar American management, Robert Hayes and William Abernathy provide ample evidence of how the short-term bias of such criteria have caused us to "manage our way to economic decline."[29]

On yet another front corporations tried to make R&D more efficient by subjecting it to a battery of new scientific management techniques, or "management science," as it was called in the parlance of the day. Management systems like the Critical Path Method (CPM) and Program Evaluation and Review Technique (PERT), originally developed for the Pentagon and NASA, were deployed to move projects through the pipeline more quickly.[30] But such solutions could not work. The new "scientific" models merely formalized and ratified the assembly-line model when what was needed was a complete break with it.

In the end both activities suffered. R&D was given short shrift, and a huge number of organizational blockages developed at the various stages of the development pipeline. The path from R&D to manufacturing became a bureaucratic nightmare of "it's not my job," "that isn't my department," and "we can't do that." It became harder to get R&D projects going, more troublesome to complete them once they were started, and exceedingly difficult to transform innovations into products.

The Separation of Innovation and Production

The 1960s and 1970s witnessed a growing physical separation of R&D from manufacturing—the sites of innovation and production, respectively. Given the prevailing specialized model of corporate organization, most companies did not feel it essential that the R&D scientists who produce inventions interact with the shop-floor workers who actually make the firm's products. On the contrary, it was believed that separating R&D labs from factories would provide researchers with the insulated, ivory-tower environment needed to achieve technical progress, far away from the noise, pollution, and hubbub of the factory floor. In the words of one former director of an industrial research lab: "Our approach to research in 1965 was much like that of other companies. Research was somewhat of an island unto itself; there was very little interaction with other company functions."[31]

With such beliefs in hand, our largest corporations developed a new strategy for locating various activities. Basically, different corporate functions—R&D, manufacturing, and administration—would be parceled out and located in the most advantageous places.[32] Manufacturing was relocated to the American Sun Belt and later to the Third World in order to cut wage costs and avoid unions. R&D was moved to suburban campuses near where most R&D workers lived, while high-level management and administrative functions were placed either in "redeveloped" downtown business districts or in suburban office complexes. Geographers and regional development specialists have advanced the concept of a "new spatial division of labor" to capture this new corporate geography.[33]

All of the elements of this changing corporate geography are evident in the locational choices made by Westinghouse. In the early years of the twentieth century the company established its original R&D lab on the site of its sprawling East Pittsburgh manufacturing center. As it expanded in the 1950s and 1960s, various product lines were relocated from East Pittsburgh to new factories in other parts of the world. Eventually, it no longer seemed necessary to keep the R&D lab near the East Pittsburgh site, which accounted for a declining share of output. In the 1960s the decision was made to move R&D to a new suburban campus miles from the old plant. This seemed to make sense, since the new central R&D facility was supposed to serve the entire company. Now that the East Pittsburgh plant has been shuttered, Westinghouse's R&D facility is almost completely isolated from its manufacturing operations. In fact, there are rumors that Westinghouse may move its R&D lab out of the Pittsburgh region in the next few years.[34]

The Costs of Separating Innovation and Production

As the gulf between the sites of innovation and production widened, it became more difficult to turn innovations into mass-produced products. Research became its own little world, oriented toward technological breakthroughs and separated physically, socially, and psychologically from the manufacturing divisions of the modern corporation. A sense of supremacy emerged in R&D laboratories, among researchers who seldom if ever stepped onto the dirty, noisy factory floor. Divorced from production, R&D contributed little in the way of incremental improvements in products or processes.

R&D came increasingly to be viewed as an expensive but necessary gamble, in which the costs of countless losses could be covered by one big "home run." RCA's central lab, for example, continued to generate important new technology in which neither management nor its manufacturing divisions had much interest.[35] Beset with an abundance of technologies they could not commercialize, RCA executives decided to make money by selling them off to foreign companies. In other words, R&D had become so disconnected from the day-to-day activities of the corporations that management decided to use its discoveries as a commodity to be sold to other companies.

Manufacturing divisions, in turn, distanced themselves even further from R&D labs. They paid less and less attention to R&D, which no longer even attempted to deliver much-needed incremental product or process improvements. In fact, manufacturing divisions were forced to develop their own capacity to make these kinds of innovations. As Margaret Graham so aptly puts it: "More and more heads of manufacturing were evaluated as profit centers and were only interested in short term product or process research support that would improve their immediate earnings picture."[36] Those in manufacturing came to view R&D labs as corporate frivolities irrelevant to the day-to-day needs of actual production, and they began to resent the better pay and privileged status of R&D workers. At RCA even scientists and engineers in manufacturing units shared this general resentment of the better conditions, pay, and prestige accorded to researchers at the central lab, or "country club," as it came to be called—"a place where RCA resources were squandered on exotic or impractical ideas."[37]

The consequences of all this were both profound and disastrous. The connections between R&D and production were irrevocably severed. New ideas and inventions were stranded in a "twilight zone" between R&D and production. American industry went from a system in which innovation and production were closely linked to one in which it became increasingly difficult to produce the research labs' developments economically.

Again, corporate solutions only complicated the original problems. Some companies, like DuPont, responded to the growing gulf between the R&D and manufacturing by creating internal "venture" divisions designed to turn promising R&D into new products or in some cases into new businesses.[38] But few of these new venture divisions proved successful. For example, none of DuPont's major new internal ventures or spin-off companies amounted to much. The reason for this was basic: new venture divisions simply added another intermediate level to an already overblown and unwieldy R&D bureaucracy.[39] Here again, large corporations showed that they were oblivious to the need for more fundamental kinds of restructuring—the creation of new mechanisms to link research and manufacturing by locating these activities closer together, setting up interorganizational teams, or even rotating people.

Xerox PARC: Separation in the Extreme

Xerox Palo Alto Research Center (PARC) is the classic case of what happens when R&D and manufacturing are physically and socially isolated from one another. PARC was the source of some of the most important advances in computer technology in the past three decades—virtually none of which were exploited by Xerox. This failure is especially illustrative, since Xerox's pioneering strides in photocopying made it one of America's quintessential follow-through companies. The story of Xerox PARC is an illuminating case study of the dangers of separating innovation and production.[40]

PARC was launched in 1970 to develop innovative new products that would realize Xerox's vision of the "office of the future." PARC quickly invented a cornucopia of advanced products, including the basic operating system and software to create a user-friendly Macintosh-like computer, stand-alone engineering workstations, the computer mouse, computer networking, and laser printing, among many others.

But from the outset PARC's mission within Xerox was somewhat cloudy. Initially, there were significant internal debates over where to locate PARC. While some argued for a location close to headquarters, others argued for a location near the center of the computer industry, Silicon Valley. A further complication was that some members of top management envisioned PARC as a support operation for Xerox's recent computer acquisition, Scientific Data Systems. However, PARC's director, George Pake, and his staff had different ideas. For them, PARC represented an opportunity to build the most creative office automation development group in the world.

While Pake and his PARC staff succeeded, Xerox management was never able to grasp the importance of the cutting-edge technology developed by PARC. Management attention was legitimately deflected by Xerox's ongoing battle with a variety of Japanese copier companies, and it missed numerous opportunities to outflank competitors with the superior technology coming out of PARC.[41] PARC researchers tried repeatedly to interest top management in their inventions. Management refused to back a whole host of PARC inventions: a new Alto computer workstation, the Notetaker personal computer, a Japanese-language word processor, new graphics technologies, networking software, and numerous other technologies. One PARC researcher recounts his frustration at trying to convince Xerox management to develop the Notetaker: "The last year before I left PARC [was] spent flying around the country talking to Xerox executives, carrying Notetaker with me. Xerox executives made all sorts of promises: 'we'll buy 20,000, just talk to this executive in Virginia, then talk to this executive in Connecticut.' The company was so spread out, they never got the meeting together. After a year, I was ready to give up."[42]

Management stymied PARC scientists in their limited, futile attempts to develop prototypes and products. In one case, management continued to support the Star computer—an inferior product developed in the company's central product development unit.[43] In response, PARC researchers produced Altos surreptitiously and test marketed them to the U.S. government and various universities. Almost unbelievably, the decision to develop a commercial laser printer came only after PARC itself had begun manufacturing prototypes.

Xerox's managerial paralysis created plenty of opportunities for PARC employees to launch their own start-ups—its Ethernet networking system was commercialized by a venture capital–financed start-up company, 3Com Inc.—or for enterprising competitors to bring out their own versions, such as the Apple Macintosh, of PARC's inventions. The comments by Steven Jobs to John Sculley describing his reactions upon visiting PARC are illustrative: "I went to Xerox PARC and saw that they had all the great people and they were doing all the great things, and they just didn't see. I believe in great products and they haven't built great computer products with their technology. I just couldn't control myself."[44] Ultimately, PARC would function as a generic development laboratory for Silicon Valley, enriching entrepreneurs and venture capitalists far more than Xerox.

As the case of PARC illustrates, many of our largest, most technically advanced corporations were victims of a new kind of organizational paralysis. Blockages impeded the transformation of new inventions into actual products. Research and manufacturing were now completely separated from one another. And their separation produced two distinct mind-sets or worldviews that were unable to comprehend or communicate with one another. The breakthroughs came, but the large corporations were unable to exploit them.

Defense Spending and the
"Decommercialization" of R&D

The rise of a defense economy and the increasing military orientation of R&D in large companies was the final element in the loss of follow-through.[45] Before World War II the Department of Agriculture actually supported more research than did the military or any other federal agency. Mobilization for war brought a tremendous expansion in military-related research, from $48 million in 1940 to $500 million in 1945.[46] The contribution of science and technology to the war effort created a new constituency for industrial R&D: the military. Dwight D. Eisenhower succinctly explained the increasing importance of industrial research to the military: "The armed forces could not have won the war alone. Scientists and businessmen contributed techniques and weapons which enabled us to outwit and overwhelm the enemy. Their understanding of the Army's needs made possible the highest degree of cooperation. This pattern of integration must be translated into a peacetime counterpart."[47]

Defense spending grew massively during the cold war years of the 1950s and 1960s, as the U.S. became a "permanent war economy."[48] In 1964 defense spending for R&D hit $7.5 billion, more than half of total R&D spending. By the early 1980s, defense-related R&D expenditures, which include defense, space, and atomic energy funds, made up more than three quarters of the federal research budget.[49] Over the entire postwar period from 1947 to 1987, defense expenditures totaled more than $7.6 trillion—a sum that is roughly equal to the total value of all U.S. plant and equipment plus the entire value of the public works infrastructure.[50]

The huge captive market for defense technology led companies to shift their R&D from commercial to military technologies, which caused a massive "decommercialization" of industrial R&D. According to George Wise, by 1950 nearly half of GE's industrial R&D was funded by the Pentagon; and similar percentages were common at other industrial corporations.[51] Defense R&D was the premier safe bet, especially since R&D budgets could be padded easily through cost overruns and gold-plating. The defense and commercial sectors of the economy came to be characterized by separate and distinct economic incentives: the former to develop increasingly complex hardware to satisfy the needs of specific Pentagon projects regardless of cost, the latter simply to make better, cheaper products. Some companies, like IBM, actually segregated commercial and defense R&D to prevent military practices from "infecting" commercial activities.

Profits from military projects were not only assured, they were higher than those of the commercial market. Companies that chose the protective shell of the military "grants economy" could neglect commercial innovation and still increase their profits. Moreover, the competition for lucrative military contracts caused companies to use a large and increasing share of commercially generated resources to gear up to win lucrative defense awards. Reuben Mettler, president of TRW, the giant defense contractor, describes this aspect of the defense R&D system: "[T]he competitive aspect is to win the contract up front. It's a Russian-roulette type of competition. There aren't very many contracts, the stakes are large and the initial research investment is enormous."[52] Once the contract is won, the winner no longer is concerned with increasing efficiency. This is entirely different from the constant improvement necessary to achieve success in the civilian economy.

The cozy relationship between the Defense Department and American corporations caused many firms and industries to become complacent and ultimately to fall behind more aggressive foreign competitors. This was a critical factor in the decline of the American machine tool industry where military funding led to the development of computerized numerical control technology, which later proved inappropriate for commercial use.[53] Defense spending also caused a number of our electronics giants to fall far behind their competitors in the production of commercial semiconductors, a problem that has now come back to haunt American high technology.[54]

While supporters of the Defense Department have argued that military spending has produced important civilian, or dual-use, technologies, the evidence suggests that this is not the case. With the exception of the aircraft and aerospace industries, involvement in defense spending generated very few dual-use technologies, or civilian spillovers, in the postwar years.[55] And as military technologies became more advanced and exotic, they became less and less useful for commercial purposes. One study of the development of semiconductor technology concluded that:

> none of the major innovations in semiconductors have been a direct result of defense sponsored projects. Major advances in semiconductor technology have with few exceptions been developed and patented by firms or individuals without government research funding. Far fewer patents have resulted from defense supported R&D than from commercially funded R&D, and a far smaller proportion of those which have resulted from defense support have had any commercial use. . . . Those transfers which have occurred . . . have required essentially new designs for commercial applications.[56]

A broader retrospective study of the postwar era conducted by the National Academy of Engineering in 1974 came to a similar conclusion, finding little evidence of commercial spillovers from defense technology:

> With few exceptions, the vast technology developed by federally funded [military] programs since World War II has not resulted in widespread "spinoffs" of secondary or additional applications of practical products, processes and services that have made an impact on the nation's economic growth, industrial productivity, employment gains and foreign trade.[57]

Rather than enhancing commercial efforts, massive amounts of military R&D tended to crowd out civilian R&D. According to Frank Lichtenberg, who has written extensively on the subject, defense funding drained a huge amount of capital and deflected a tremendous number of scientists and engineers away from commercial technology.[58] In other words, defense spending created a huge "opportunity cost" for the rest of the U.S. economy. When U.S. commercial producers competed with defense sector firms for scarce resources and talent, they were simply unable to offer the kinds of salaries that a cost-plus defense contractor could offer. The result: the best and the brightest of our engineering talent went to work for the military, leaving the "leftovers" to work on commercial technology.

Complicating this, the Pentagon's penchant for secrecy, classified information, and a "controlled" environment only made innovation more difficult to achieve. Scientists found it difficult to talk to one another or to collaborate on interesting projects for fear of spies and leaks. This problem was recognized quite early on by James Conant, former president of Harvard University and architect of postwar science and technology policy: "[O]ne must ponder on the consequences of the vast sums of money now being spent on secret military research and development undertakings. One cannot help wondering how long a large fraction of our scientific manpower can be employed in this atypical scientific work without threatening the traditions that have made science possible."[59]

The ascendance of the Defense Department as a funder of technology was a major contributor to the breakthrough ideology that afflicts us to this day. Breakthroughs were needed to "race" the Russians across a series of technological frontiers: the hydrogen bomb, space, the moon, and much later "Star Wars." Huge amounts of resources were poured into these military or quasi-military megaprojects, leaving very little for other areas and least of all factory production. Mass manufacturing of the sort needed to produce high-quality consumer products was neglected, because high-tech defense depended upon very specialized forms of production. Eventually, the breakthrough bias of the defense sector came to be shared by the civilian sector as well. Follow-through suffered accordingly.

A host of problems eventually cost us our earlier ability to connect R&D and production. Some of these were internal to R&D labs, some stemmed from the changing management structure of the corporation, and still others were externally imposed. In the space of fifty years, our economy lost the world's premier institutional mechanisms for linking R&D to manufacturing and was left in a condition where these same institutional mechanisms would no longer function. The root of these problems was organizational—a deepening organizational sclerosis that large corporations and other economic and political institutions were unable to change. It is indeed astonishing that the U.S. was able to remain hegemonic for so long.

All of this brings us to a very basic point. The linkage between R&D and manufacturing requires a delicate balance. If R&D is too tightly bound to manufacturing, it may become overwhelmed with the mundane work of developing minor improvements and therefore neglect exploratory research, one of its fundamental missions. But when the two critical components of R&D and manufacturing begin to grow too far apart, it may become increasingly difficult to harness and utilize major technical advances.

Although a technological revolution was gestating inside our large companies and in other U.S. research institutions, these institutions were unable to give birth to the new possibilities. Yet U.S. capitalism is a dynamic system. Even as the old system frustrated and stultified its employees, a new set of economic institutions emerged to provide an outlet for a few of the most creative individuals. A parallel economy based on these breakthroughs grew up in the shadows—an answer but not a solution for America's move into the high-technology age.

3

A New Breakthrough System
for High Technology

■

There is a basic incompatibility of the inventor and the large corporation.
—JACK KILBY, co-inventor of the integrated circuit.[1]

Having an unstructured environment promotes creativity and team work through increased interaction between individuals and groups.
—Tandem Computers, Inc. brochure[2]

The new breakthrough economy was a response to the limits and weaknesses of the old follow-through economy. It arose mainly to take advantage of highly profitable growth areas that large Fordist companies ignored or were unable to take advantage of. A new breed of high-technology industrialists—a new group of historical subjects— saw the potential of these new technologies and stepped forward to create the new organizations and institutional structures needed to cross the extraordinary technological frontiers and exploit the new industrial opportunities of the high-technology age.

Although the details vary from technology to technology and industry to industry, the basic message remains the same: the large companies failed to move into new technological openings, leaving the door open for the entrepreneurial start-ups of the breakthrough economy.

Consider the birth of the computer industry. Large companies, including IBM, were painfully slow to move into important new technological openings, even though many had the technological capabilities needed to do so. In fact, the first commercial computers were built by start-up companies: Eckert-Mauchly Corporation and Engineering Research Associates (ERA). The latter controlled some 80 percent of the computer market in 1952.[3] Both companies suffered chronic financing difficulties, which were no doubt exacerbated by the absence of a formal venture capital industry, and both eventually sold out to Remington Rand. IBM entered later, in the mid-1950s, with a series of products that initially brought mass production (its 701 and 650 lines) and later compatibility (the famous 360) to the computer industry.[4] Following IBM's lead, a host of large companies—RCA, GE, NCR, Burroughs—entered the commercial computer business in the late 1950s and early 1960s.

Nevertheless, much of the industry's technological evolution, especially the creation of whole new branches of computer technology, has been driven by entrepreneurial start-up companies. Digital Equipment Corporation (DEC), the pioneer in minicomputers, was founded in 1957 by Kenneth Olsen and Harlan Anderson of MIT's Lincoln Laboratories and was backed by the country's first institutional venture fund, American Research and Development (ARD).[5] Control Data Corporation (CDC), another 1957 start-up, was launched by ERA alumnus William Norris and introduced high-end scientific computers, the ancestors of supercomputers.[6] Later companies like Tandem (fault-tolerant computers), Apple (personal computers), and Sun Microsystems (engineering workstations) opened up other branches of the computer industry. Large companies—even IBM, the most successful computer firm of all time—were unable to anticipate these major new openings, leaving the field wide open for exploitation by start-ups. In fact, over the past two decades many large U.S. companies have been driven out of the computer industry: GE and RCA have abandoned computers, Burroughs and Sperry have merged to become Unisys, and Honeywell has essentially become a distributor for Japanese computers.

Next, consider semiconductors, where large companies did little to capitalize on sizable "first mover" advantages. The transistor was invented in 1947 by three Bell Labs scientists, William Shockley, Walter Brattain, and John Bardeen. In the immediate postwar period companies such as GE, RCA, AT&T, and Raytheon were the undisputed technological leaders in semiconductor technology. But these established companies did not reap the rewards of that lead. AT&T was constrained by a Justice Department ruling that prohibited it from selling semiconductors and, more importantly, by its own emphasis on defense-related semiconductors.[7] Interested in developing real products, many of AT&T's top semiconductor scientists left the company. In 1951 Gordon Teal left Bell to take over the semiconductor division of Texas Instruments, a small company that had previously specialized in geophysical exploration. Three years later William Shockley left to form Shockley Semiconductor in Palo Alto, California.[8]

In 1957 eight employees left Shockley to form Fairchild. In what has now become a high-tech legend, the eight sketched out a crude business plan and contacted Arthur Rock, who arranged financing from the East Coast firm Fairchild Camera and Instrument. Fairchild emerged almost immediately as a leader in the rapidly expanding semiconductor industry. In 1958 one of its scientists, Jean Hoerni, invented the planar process that made mass production of semiconductors possible.[9] In 1959 another of the Fairchild's founders, Robert Noyce, coinvented the integrated circuit. And in 1961 its cutting-edge R&D division developed the bipolar circuit. Fairchild grew quickly, outflanking electronics giants GE, RCA, Philco, Westinghouse, Sylvania, Raytheon, and Hughes.[10] Fairchild's dramatic success exerted a powerful "demonstration effect," initiating a trend of entrepreneurial innovation in semiconductors. From this point onward, the U.S. semiconductor industry was driven by a seemingly endless cycle of start-ups.

Last, consider biotechnology. Here again, large corporations were laggard, leaving the door open for venture capital–backed start-ups. In fact, pharmaceutical and chemical companies developed a myopic outlook that prevented them from seeing the commercial potential of scientific advances in molecular biology. Peter Farley, one of the founders of Cetus, an early biotechnology start-up, recalls that in the early and mid 1970s "we were thrown out of some of the finer offices in the chemical and pharmaceutical industries. This was before recombinant DNA: but recombinant DNA was absolutely predicted by our scientific advisors."[11] Large companies hesitated as major new discoveries in molecular biology opened a window of opportunity for new players. Venture capitalists moved into this gap quickly, financing small start-ups launched mainly by university professors.

A new start-up company, Genentech, led the biotechnology "gold rush" of the late 1970s. It was founded by the venture capitalist Robert Swanson, who had managed a previous investment in Cetus, and Herbert Boyer, a Nobel prize–winning molecular biologist at the University of California, San Francisco. Together, Swanson and Boyer drew up a business plan detailing the steps necessary to make biotechnology a commercial reality. They then approached Thomas Perkins, Swanson's former boss at the venture capital fund Kleiner Perkins, and convinced him to invest $100,000 in the venture.[12] Based largely upon its promise to produce commercial biotechnology products, Genentech's initial public stock offering was so oversubscribed that its stock price soared from $35 to a high of $89 the day it was issued (the price quickly settled in the low forties). This provided ample proof of the commercial possibilities of biotechnology and motivated venture capitalists to back hundreds of new biotechnology start-ups.

By the mid-1970s the high-technology start-ups of the new breakthrough economy had given rise to three of the most important technologies of the late twentieth century. Their rapid advance over the large resource-rich corporations of the old follow-through economy was nothing less than astounding. What was the key to their success? The answer lies in the new model of high-technology industrial organization they created.

Inside the Hothouse

Teams are the cornerstone of this new model of organization, replacing the extreme functional specialization and managerial sclerosis of large Fordist companies with work environments that are decentralized, nonhierarchical, and integrated.[13] Teams enhance breakthrough capacity by creating an environment that unleashes the creativity and intelligence of high-technology think-workers who are encouraged to interact and jointly develop ideas.* Leslie Misrock, a well-known lawyer for the biotechnology industry, explains: "Real innovation comes out of the hothouse atmosphere at . . . small companies, but is somehow stifled at large companies."[14]

Teams overcome task specialization with a collective group approach to problem solving. They give everyone a real piece of the action and allow them to see the ultimate fruits of their labor. *The Soul of a New Machine,* Tracy Kidder's compelling account of the design of Data General's Eclipse minicomputer, provides important insights into this new organizational environment:

> The work was divided but it was not cut to ribbons. Everyone got responsibility for some important part of the machine, many got to choose their piece, and each portion required more than routine labor. The unspoken rules of the group were Darwinian, but many of those who made it through [the life of the project] declared that they had been given as much freedom as they could have wished for.[15]

In pioneering companies such as Fairchild and DEC, teams became a conscious alternative to the extreme specialization of the assembly-line of innovation. The university served in part as a model, but the emphasis was on practical or commercial research. From the outset R&D teams were designed to encourage initiative and creativity.

Team membership has an important motivational aspect. Small teams enable think-workers to devote themselves to one project, which gives workers the feeling that they can control their own destiny. They generate unique internal bonds that motivate their members to work exceptionally hard. The mutual dependence of the team environment generates the high level of commitment needed to take ideas from the drawing board to commercial production in amazingly short time frames. It is a curious anomaly that while start-ups emphasize the importance of the "lone genius," the reality is one of teams of employees working to develop the new high-technology products.

---

*An ideal term for the engineers and R&D scientists of the high technology age has not yet been coined. We advance the term "think-workers" to capture the fundamental fact that their work has almost nothing to do with their muscles but rather relies upon their mental facilities.

Small teams can, and often do, beat large R&D bureaucracies to a major discovery, since in many cases the addition of more people only slows down R&D efforts.[16] David Lundstrom, an original UNIVAC engineer, makes this point very clearly: "[T]he most successful computer developments are those done by small, intensely focused teams, working in the same facility or, better still, in the same room, and under tight schedule pressure. Amazingly, the time constraints seem to produce designs that are cleaner, not sloppier."[17]

R&D teams are in many respects self-organizing—members volunteer rather than waiting to be assigned. This in itself generates higher levels of commitment and motivation. An example of how one engineer became a member of Apple's highly innovative Macintosh team is instructive: "Burrell Smith, untrained in computer mechanics, [came] out to California and [got] a job in Apple's repair department. Then the Mac [got] going and he took advantage of the freedom Apple gave its employees to insinuate himself into the process, mostly by hanging around. Back in the early days, we did . . . whatever we wanted to do. There were no rules."[18]

In addition to a heavy reliance on teams, many firms have developed an informal process of rotation, as scientists and engineers are switched to different positions. The small size of these companies often makes such switches necessary, because there are not enough people to do everything that needs to be done, especially as the company grows and new needs develop. A few larger companies like Hewlett-Packard have formalized this process by strongly encouraging or even making it mandatory for employees to change jobs regularly and rotate to new positions. An HP employee describes the company's "career maze" as follows: "In my first eight years, I guess I had about seven or eight different jobs and four different functions. A couple were lateral moves, some of them were promotions. It's not necessary to get a bigger title or jump to a new level in the pay system to be given more responsibility."[19]

R&D teams often operate in a situation of decentralized decision-making authority, so that scientists can do things on their own without having to wait for permission from the higher-ups. A newly hired director of research for a prominent biotechnology company told one of us that on his first day of work there was no one to tell him what to do; he was simply urged "to do what was necessary and be creative."[20] This is a major advance over the large corporations of the old follow-through economy in which decision-making authority was lodged high in the managerial stratosphere. Intel, for example, "encourages decision-making at the lowest possible levels. Employees are expected to face up to difficult decisions whether they be business, organizational or personal. . . . Intel expects its people to take on a task and deal with it without being told."[21]

Intel thus recognizes a simple fact of the R&D process: the highly irregular work process of R&D scientists makes it difficult to manage them according to traditional bureaucratic principles. An R&D worker who appears to be doing little may be immersed in the kinds of thinking and creativity it takes to make a major innovation, while another may spend his time busily working at a lab bench and come up with nothing.

The decentralization of decision making is a powerful method for tapping the knowledge of R&D scientists directly. It leaves responsibility for monitoring the process with the people who know best—actual R&D workers. Those with direct practical knowledge are encouraged to adapt and change technologies unencumbered by a maze of check-offs and approvals. And of course, decisions are more likely to be followed when they are made by those who execute them.

To strengthen and reinforce decentralized decision making, many high-technology companies have established decentralized funding systems that channel funds to "extracurricular" projects. Employees can secure seed funding for new ideas and exploratory projects. On a smaller scale, some companies, like Hewlett-Packard, allow engineers to use company resources to work on their own ideas, and they encourage employees to take supplies and instruments home for garage tinkering in the hope that this will increase their productivity.[22] To some extent then, think-workers are afforded the freedom to pursue their own goals and call it work. The technological vitality of the breakthrough economy derives in large part from the combination of R&D teams and decentralized decision making, which opens up unforeseen channels of information and ensures that high-technology companies tap much more of the intelligence and creativity of think-workers.

Pumping Work Out of High-Tech
Think-Workers

An important reason for the success of the high-technology start-ups is that employees are willing to put in superhuman hours—a work-week of seventy, eighty, or ninety hours is not unusual. Apple's Macintosh R&D group "[w]ore sweat-shirts inscribed '90 hours a week and loving it.' If someone spotted a programmer early in the morning, it meant he'd been up working all night. One ritual was fresh fritters from the Doughnut Wheel at 4 A.M."[23] The pressure cooker environment of high-technology start-ups is amazingly effective at pumping work out of think-workers, who work longer, harder, and more furiously than did the R&D professionals of the old follow-through economy. Long hours and hard work are an important part of the ability of start-up companies to quickly generate new break-throughs.

There are three basic mechanisms for eliciting this superhuman work effort: (1) remuneration schemes that link individual rewards directly to the success of the enterprise, (2) recruitment devices that locate highly motivated workers, and (3) management techniques that push workers to the limit.

The equity stock ownership plans offered by high-technology companies motivate work by creating an identity of sorts between individual and corporate interests, tying personal gain to corporate performance. This type of remuneration is a far more effective motivator than straight paychecks. According to Jerry Sanders of AMD, incentive pay schemes "tie the improvement of the employees' economic condition to the improvement of the corporate economic condition in such a way that they understand that when the good times come they participate."[24]

Stock ownership provides a concrete mechanism for allowing individuals to profit from the fruits of their labor. Nancy Dorfman believes that "small firms have an advantage in acquiring the talented scientific, technical and managerial personnel needed to produce innovations. . . . [E]stablished corporations can rarely offer potential innovators the financial rewards of genuine participation in a successful innovative undertaking that a small, new enterprises can provide."[25] The attractivenesss of these schemes is dramatically enhanced by the "big scores" achieved by other start-ups. For example, when Apple's stock surged in June 1983, roughly one hundred employees became millionaires.[26] Thus, employee stock ownership provides a method by which certain elite employees are able to enrich themselves as new breakthroughs succeed.

While the roots of employee stock ownership go back a long way, it was Fairchild that pioneered the use of equity ownership for high-technology think-workers. According to Fairchild cofounder Robert Noyce: "Fairchild was the first time that scientists and technologists really got themselves in the position of controlling the operation, with high financial rewards for successful experimentation. We soon had a policy of spreading those rewards through the ranks and pretty soon Fairchild became the premier semiconductor laboratory in the world."[27]

Fairchild's founding agreement allowed its corporate parent, Fairchild Camera, to buy the company stock from its founders at a prearranged price. This angered the founders, and when they left Fairchild they did not repeat the mistake of giving an investor the right to buy the company out from under its founders. Eugene Kleiner, the leading venture capitalist and one of Fairchild's founding eight observes:

> When Fairchild financed us, they gave us 100 percent ownership of the company. But they reserved the right to buy the company back at a predetermined price. Once they exercised that option, once they bought back the company, they were slow to give out options. That's one reason people left. That was also about the time John Wilson [of what would become an important Palo Alto law firm Wilson Sonsini, which specialized in legal work for new firms] began developing plans to give entrepreneurs ownership.[28]

When Robert Noyce and Gordon Moore left Fairchild to form Intel, they devised an extensive plan of employee stock ownership considered by many to be the model for high-technology equity incentive plans.

However, employee stock ownership is effective only until the stock vests. Stock option plans with four- or five-year vesting schedules are commonly called "golden handcuffs," because they temporarily tie workers to a company. Lotus, for example, lost a large number of employees in July 1988, after their stock options had vested.[29] (We return to the issue of labor mobility in chapter 5.)

Equity ownership is also used to motivate lower-level employees. Although significant stock ownership typically extends only to key employees, a number of companies distribute smaller amounts of stock to middle-level researchers and managers. James Treybig, president of Tandem Computers, refers to Tandem as a "socialist company," because every employee is given a certain amount of stock.[30] National Semiconductor operates what it calls a "paysop"; that is, stock is distributed in the form of bonuses that accrue equally to all employees. Even when the amount of stock is too small to provide economic motivation, it can perform an important psychological function because employees perceive themselves to be partial owners of the company. Profit sharing is another type of employee incentive program that is commonly used to motivate lower-level employees. Like stock ownership, profit sharing ties the employees' financial condition to that of the company. Of course, compensation schemes vary according to the stage of growth of the company. For example, stock ownership is used more heavily during the initial start-up phase, whereas other incentives such as profit sharing are more common at later stages.[31]

Selecting the "Right Kind" of People

But there is more to it than greed. High-technology companies actively target their recruitment at self-motivated people who will sacrifice to achieve technological and financial goals. The ability to work long hours is an absolutely essential hiring criteria. One candidate for a Data General development team was rejected because he listed family life as one of his avocations. This was considered "evidence" that he might not be able to meet the demands of the job.[32] Finding highly motivated people is critical to a small company with only a handful of employees on a breakthrough-oriented project team. As one high-technology executive put it: "What makes a high-technology company is not technology. Technology is perishable, it changes too fast and needs to be constantly regenerated. The only way to regenerate it is to have the best people in the business working for you."[33]

The recruitment process is designed to select employees who will work hard and fit into a team environment. New recruits are judged not only according to standards of competence and professional ability but according to their ability to fit into the social context of a project team. This is why so many high-tech employees are recruited via the established networks of current employees. Potential employees may spend as much as two or three days meeting with team members and management. Research scientists are typically interviewed by personnel officials, by all the researchers in the group in which they will work, and by researchers in other groups as well. Top management actively participates in the recruitment of key people. At Tandem Computers, a company with revenues of over $1 billion in 1989, the president still participates in two or three job interviews per week.[34]

The penalty for recruiting the wrong person can be severe. It can mean a disrupted R&D project and a drastic slowdown in work effort. Termination of a problematic key employee can be traumatic for the work environment in a small company. It can also be very expensive, since the employee has usually been provided with stock options and perhaps a signing bonus.

Sometimes the methods used to identify suitable personnel are offbeat but fascinating for their insight into the social relations of high-technology firms. The director of human resources for Borland International has a green plastic frog on her desk. The frog hops when a small bulb is squeezed. If a potential employee grabs it and examines it, she knows she has a "winner."[35] Janet Axelrod, then vice president of human resources for Lotus, would ask potential employees their opinion of Fidel Castro, whom she admires, in order to gauge their ability to fit into Lotus's progressive social environment.[36] The Apple Macintosh team looked very favorably upon job candidates who played Space Invader well, because team members liked to play the game.

Motivation Is More Than Money

The characteristics of the organization itself generate additional motivation. As we have seen, teams generate a sense of camaraderie and mutual dependence that can spur intense work effort. And many high-technology companies allow R&D workers to create their own work environments—a policy designed at least in part to generate maximum work effort. Consider the motivational effects of the Apple's Macintosh product development center:

> An expensive, high-tech stereo system blasted the Pointer Sisters' "I'm So Excited" though 6-foot-tall, slim electrostatic speakers, Steve [Jobs] had ordered every single compact digital disc available, . . . and the music was as constant as the air conditioning. . . . [S]omeone else would play the concert-level Bosendorfer piano in the corner. Parked against one wall was Steve's motorcycle. A little Heathkit robot would scamper out in the hall from the software room. The scene looked more like a college rec room than a corporate product development center.[37]

Work effort is also motivated by the fact that high-technology think-workers are producing whole products, not just parts of products. These think-workers can rest assured that their new products will pass or fail the test of the market rather than be killed in the corporate bureaucracy. Peggy Asprey, a cofounder of David Systems, a Silicon Valley communications company, explains: "In a startup . . . whatever you do, if you make it, it's going to be out there and sold. At Bell Northern Research I was on a project; we were working 80 hours a week, literally. And when we were done they canceled the project. What have I got? Nothing, nothing."[38]

Of course, think-workers are often motivated by the excitement and content of their work. Lee Felsenstein describes his work hours while designing the Osborne Computer: "I worked all three shifts on some days. The difference between tools and toys is not much."[39]

High-technology companies motivate work by manipulating the status differentials between different types of workers. Engineers are frequently given status as high or higher than that given managers. Parallel career tracks are common, so engineers do not have to become managers in order to succeed financially.[40] However, it should be noted that these technologies are changing so rapidly that the technical skills of engineers more than ten years out of college are often obsolete. In that case, they may be encouraged to move onto the management track.

The culture and reward system developed by many companies encourages intense work effort. Some companies award prizes to employees who have made special contributions. Others, like Rolm and Hewlett-Packard, offer paid sabbaticals to veteran employees who have made important contributions to the company. On a much smaller scale, numerous Silicon Valley companies have Friday afternoon "beer busts" to reward employees for grueling workweeks and to get them up for working weekends. These events also have the side benefit of fostering unity, while creating an "unstructured" environment for informal information exchange and discussion.

Making R&D Relevant

Breakthrough capacity is further reinforced by the highly focused nature of R&D. The high-technology companies of the breakthrough economy organize R&D in a way that emphasizes development over research. They devote a much larger share of resources to R&D than do the large corporations of the follow-through economy. Most of these resources are channeled into developing actual product technologies.

A 1989 study found that new Silicon Valley companies spend nearly 350 percent of sales on R&D during the first year of existence—an unbelievably large commitment to R&D.[41] According to recent rankings, venture capital–backed start-ups dominate the top hundred R&D spenders in microelectronics, as measured by percentage of sales invested in R&D. The highest ranking large company was DEC (an earlier start-up), in twenty-sixth place. The next were Hewlett-Packard in thirty-fifth place, followed by IBM in thirty-seventh place, Xerox in sixty-fourth place, GM-Hughes Electronics in eighty-third place, Honeywell in eighty-eighth place, and Zenith, which ranked ninety-sixth. Venture capital–backed start-ups such as Chips and Technologies, Ashton-Tate, Lotus, Microsoft, Cray Research, Sun Microsystems, and Tandem Computers comprised the entire top fifteen companies in terms of R&D spending per employee. These companies devote between $20,000 and $60,000 per employee to R&D. According to recent statistics from the National Science Foundation, small companies (with 50 employees or less) increased their share of total corporate R&D spending from 6 percent in 1980 to 12 percent in 1987.[42]

A variety of organizational factors work to keep R&D focused on commercial products. High-technology companies go to great pains to eliminate the invisible walls that divide R&D from other corporate activities. R&D and management are typically housed in the same building, with laboratory facilities adjacent to management offices. Flatter organizational pyramids mean that R&D scientists and corporate managers typically know one another by name and are used to cooperating with one another. Avoidance of an overblown bureaucracy means that projects do not get lost in the stratosphere of a managerial bureaucracy.

"Management by walking around" keeps managers in close contact with R&D scientists. Many high-technology companies have an "open door" policy, which means that researchers are free to drop in and express their ideas to managers. Charles Sporck explains: "There is a real functional and symbolic value in the absence of walls and doors—it is easier to talk to your neighbor or to your boss. The threshold for cross communication is very low. Ideas or complaints don't have to wait long to get aired."[43] Such practices increase the chance of serendipitous contacts that can lead to ongoing interchanges, new ideas, and ultimately new innovations.

Decision making by consensus is another technique that is used by such companies as Sun Microsystems and Compaq Computer to foster interaction between R&D workers and management. Scott McNealy, president of Sun Microsystems, enunciates this philosophy: "Give me a draw and I'll make the decision. But I wouldn't edict something on an 11-to-1 vote."[44] The objective is to ensure that innovations are turned into commercial products as quickly as possible. Management's role is to keep the process moving along at the fastest possible pace and to eliminate any blockages that might bog it down.

The use of "network" organization also helps keep R&D connected to commercial objectives. In the 1970s, for example, Intel organized interdepartmental councils comprised of engineering, marketing, and other managers who meet regularly to review progress and set new objectives. Advanced Micro Devices links various activities through similar councils, such as its "council for computer-aided design." Basically, corporate organization is designed to maximize interaction and communication between R&D scientists, marketing, and high-level corporate managers.

Other firms believe that smallness is essential to retain an entrepreneurial spirit, keep people motivated, and make sure R&D is in sync with commercial objectives. To do this, they hive off divisions and create "companies within companies." Corporate headquarters is conceived of as playing the role of venture capitalist, providing the initial infusion of start-up capital, after which entrepreneurial divisions are expected to go it alone. Hewlett-Packard was one of the first high-technology companies to organize "entrepreneurial divisions" with their own R&D, engineering, marketing, manufacturing, and support operations. Charles Sporck describes the use of network organization at his company: "A good way to look at National [Semiconductor] is as a cluster of related businesses, which are run by entrepreneurs with a great deal of freedom and a great deal of accountability. The corporation is an umbrella which provides resources and centralized support, but is decentralized in a way that allows us to be very responsive to market conditions."[45]

This strategy of keeping things small can generate problems, especially as companies grow. By the early 1980s, for example, Hewlett-Packard had spawned fifty-four entrepreneurial divisions, making a range of products from advanced scientific instruments to electronic calculators and computers. These divisions often did not know what their counterparts were doing; and different divisions began to pull the company in different directions. This was especially problematic in the computer area, as different divisions began to develop incompatible products. In 1984 the company responded by collapsing its fifty-four quasi-independent units under two broad product divisions. As this case illustrates, there is a tension between centralization and decentralization in high-technology management—a tension that can become a serious problem when start-up companies grow too big for traditional entrepreneurial management.[46]

Defense Funds Are Less Important

Breakthrough capacity is also reinforced by the fact that high-technology start-ups work mainly on civilian technologies and are not deflected into the defense economy. Contrary to conventional wisdom, most high-technology start-ups are not creatures of the Pentagon. In fact, while many high-technology start-ups sell products to the military, most are quite hesitant to accept defense funding for R&D.

Conscious avoidance of "defense dependency" is a major reason that the high-technology start-ups of the breakthrough economy are able to remain at the commercial technological cutting edge. They do not want defense to distract them from directly commercial R&D or to control their research agenda, and they prefer to devote scarce human resources and management talent to commercial activities. Robert Noyce stated that when he was at Fairchild, top management consciously avoided "defense dependency."

> I think the maximum we ever got in direct military support at Fairchild was four percent of our research and development budget. . . . [I]t was largely because I had worked on military projects before at Philco and I felt that it was a waste of the asset. The direction of the research was being determined by people less competent in seeing where it ought to go, and a lot of time of the researcher's themselves was spent communicating with military people through progress reports or visits or whatever. . . . [S]elling R&D to the government was like taking your venture capital and putting it in a savings account—you're not going to make any substantial gain from it.[47]

Noyce's comments are echoed by Kenneth Olsen, founder and president of DEC: "[W]e didn't want to take military contracts. We weren't pacifists; in fact, we came from a military part of MIT, but we felt it limited our efforts to become a civilian company."[48]

The role of military spending in the emergence and evolution of cutting-edge American high technology has been far less influential than the apologists for military spending have claimed. In contrast to many observers who believe that the start-up companies of Silicon Valley and the Route 128 area around Boston owe their success to the Pentagon, it is more correct to say that defense-related purchasing created a marketplace for a number of high-technology products, such as semiconductors. But in many cutting-edge fields of technology—biotechnology, workstations, laptops, and personal computers—defense spending was simply not important.

A Support Structure for Breakthrough Innovation

Breakthrough capacity is also bolstered by the environment that surrounds high-technology companies, especially in places like Silicon Valley and Route 128. The support afforded by other cutting-edge high-technology companies, a ready pool of talented engineers, managers, and R&D scientists, an abundant supply of venture capital, business service firms like law firms and accountants, and informal networks for information exchange and technology transfer makes it easier to develop new innovations and turn them into products. Luigi Mercurio, an Italian entrepreneur in Silicon Valley, uses the phrase "virtual corporation" to describe the way the Silicon Valley technology network extends the boundary of the individual firm.[49] According to high-tech marketing executive Regis McKenna: "The infrastructure of Silicon Valley is extremely complex and supportive. It involves legal advice, copyrights and license counseling, marketing counseling, management counseling, banking and various kinds of supports, investment banks, manufacturing help and subcontractors, and a myriad of other services."[50]

We refer to the networks that support and fuel breakthrough innovation in places like Silicon Valley and Route 128 as comprising a broader "social structure of innovation."[51] It is easy to see how such a social structure of innovation can help new companies, indeed new industries, emerge and grow. It allows entrepreneurs and R&D scientists to focus their efforts on innovation and not get distracted with side issues like obtaining office or laboratory space or trying to figure out the myriad details of various legal documents.[52]

The social structure of innovation is not an abstract phenomenon but a set of institutions and activities that shape everyday behavior. As one semiconductor executive explains: "I have people call me quite frequently and say, 'Hey, did you ever run into this one?' and you say, 'Yeh, seven or eight years ago. Why don't you try this, that or the other thing?' We all get calls like that."[53]

The social structure of innovation is a major source of new ideas, market openings, and information on competitors' strategies. It can even be the starting point for new companies. It is important to recognize its limits, however. As its name implies, this set of networked institutions is geared to technological innovation but *not to production,* which as we will show increasingly takes place on a national, even global, basis (see chapters 6 and 7).

The power of the social structure of innovation is reflected in the pull it can exert on entrepreneurs and technology companies located in other parts of the country, even the rest of the world. The venture capitalist David Arscott uses the term "magnet effect" to explain the way technologists and potential entrepreneurs are attracted to Silicon Valley.[54] Synopsys, a cutting-edge start-up that makes advanced software used in the design of semiconductors, was forced to move its headquarters from North Carolina's Research Triangle to a new home to be a part of Silicon Valley's dense social structure of innovation. Synopsys founder, Aart DeGeus, explains:

We started this company in North Carolina as a spin-off from GE. All five original founders were from North Carolina. We all liked the environment there; none of us really wanted to leave. But California had the management talent we needed and North Carolina didn't. California had managers, marketing talent—the whole "business side." California also had venture capital, law firms that could help us, our advisers, and most of the customers for our product. Our venture investors encouraged us to move and were pleased when we did.[55]

Taken together, the social structure of innovation and the new forms of internal organization adopted by high-technology start-ups generate a powerful breakthrough capacity—the capacity to dramatically shorten the time it takes to make path-breaking new innovations and turn them into actual products. Perhaps the most stunning evidence of this is provided by a 1989 study by two organizational sociologists who discovered that Silicon Valley semiconductor start-ups took just one year from their founding date to produce their first working prototypes, whereas counterparts in other regions of the country took well over eighteen months. What's more, Silicon Valley start-ups had their first products out the door just five months after their prototype. All told, Silicon Valley start-ups took less than a year and a half to turn their innovations into products, a full eight months faster than innovative new ventures in other parts of the country, which took more than two years to do so.[56] This rapid turnaround time is remarkable when compared with the three-, four-, and even five-year periods that large established semiconductor companies have traditionally needed to turn ideas into reality.

Thus, at the start of the 1990s, what began four decades ago as a series of small efforts and "experiments" has grown into an entire set of institutions for developing and advancing technology. The success of this new system was premised upon its ability to overcome the stagnation and rigidity of large Fordist firms and create a hothouse environment that could unleash the tremendous creativity of high-technology think-workers. This was a long evolutionary process. Methods of organization, rules of behavior, and incentives that were once seen as unusual, indeed foreign, gradually became routine and everyday—"the way things are done." And as the system grew, it attracted more and more scientists and entrepreneurs, setting in motion a powerful process of self-reinforcing growth.

However, to build a successful institutional system for the development of new technologies required one more ingredient—capital. We now turn our attention to venture capital, the new finance capital of the breakthrough economy.

4

Venture Capital and the Breakthrough Bias

■

If you have a good idea you can get $5 million in venture capital up front. But within two to five years, you must be making money with the new idea. We cannot do what a Japanese company will, to say we want to dominate a market and that we will take losses for five years to do it.

—PHIL KAUFMAN, former president of Silicon Compiler Systems, now president of Quickturn Systems[1]

The new high-technology start-ups of the breakthrough economy demanded a new form of finance capital. The main sources of capital for the old follow-through economy came from banks and other financial institutions. But those institutions were not equipped to handle the high risks of breakthrough innovation. Bank capital required a track record, collateral, and monthly repayments—terms that were virtually impossible for most entrepreneurial start-ups to meet. And while large corporations devoted a significant amount of financial resources to R&D, this was seldom available for outside efforts, least of all to employees who wanted to spin off corporate technology.

The institution that emerged to meet these needs was venture capital, the finance capital for the breakthrough economy. Venture capital differs from previous forms of capital in important respects. Venture capitalists do not lend money; rather, they take an equity or ownership stake in the companies they finance. This enables them to undertake high-risk investments while eliminating the vexing problems of collateral and monthly loan payments. They enhance the capacity of fledgling companies to innovate and get off the ground quickly by providing both capital and start-up assistance. In doing so, they provide the fuel that heats the pressure cooker environment discussed in the previous chapter. Venture capitalists thus play the dual role of financial capitalist and technological catalyst, providing the financial resources, management advice, and contacts needed to turn ideas into products.

What Do Venture Capitalists Do?

Venture capitalists identify new businesses and technologies, invest in them, and help to develop them in order to realize the superprofits that come from breakthrough innovation. In the words of Burton McMurtry, a leading Silicon Valley venture capitalist: "Venture capital is the business of developing new businesses. Venture capitalists like to start something from nothing, to stimulate and encourage innovation."[2]

Figure 4.1 depicts the various changing functions of venture capitalists over the life cycle of a "typical" high-technology start-up.

Figure 4.1   The Venture Capital Cycle

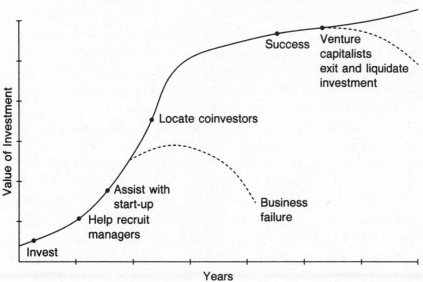

Venture capitalists depend upon well-developed social structures of innovation (discussed in chapter 3) to "create something from nothing." They use their contacts and "friends" to locate investment opportunities, put together investment syndicates, and build high-technology businesses. Regis McKenna, the marketeer and part-time venture capitalist, explains:

> The network of supporting infrastructure of Silicon Valley is the most sophisticated outside Wall Street. The catalyst for that network is the venture capital community, which has evolved to become a strategic planner, management consultant, and corporate watchdog. The network is put to work for new companies and many members of the network have been well honed on dozens of startups. . . . In fact, one of the reasons . . . many companies do succeed is because the network goes to work to help companies survive: they help them find new customers, they help them do refinancing, they help them find new managers if necessary, they help them merge with other companies to be successful.[3]

Venture capitalists depend upon their contacts and reputation among entrepreneurs and researchers who either have projects of their own or know of good projects. While they peruse all the proposals they receive, the proposals that come recommended receive the most serious consideration. If they decide a business idea is worth further consideration, venture capitalists will have long meetings with the founders and solicit information on the technical and business merits of their proposal. Members of their network as well as outside consultants often assist in this process of evaluation.

A good way to understand this referral process is as a process of "sponsored entrepreneurship."[4] The process begins when a nascent entrepreneurial group crystallizes inside an existing company. Often such a group consists of one or two engineers or R&D personnel and someone from sales or marketing and is headed or strongly backed by a well-respected and previously successful entrepreneur. The word is spread that a hot new start-up is being formed, and discussions begin between the entrepreneurial group and a number of venture capitalists from whom they seek funds. By the time the fledgling firm has produced its business plans and is ready to be launched, two or three venture capitalists are waiting in the wings to finance it.

And yet because of the inherent unpredictability of technological development, the investment decision remains based on instinct and "gut feel"—even after all the formalities of checking references and contacts and scrutinizing the investment's potential. Arthur Rock, one of the most senior and successful venture capitalists, and an initial investor in Fairchild, Intel, and Apple, sums up his motivations and goals: "I want to build great companies, that's how I get my kicks. I look for people who want to do the same thing."[5] Venture capitalists ultimately invest in only a small fraction of the proposals they review. To protect their investments, they typically seek at least 51 percent of the ownership of a company and a seat on the corporate board of directors. This position provides a vantage point from which to monitor management decisions and to pressure the company to reach various milestones quickly.

Venture capital is well suited to the demands of high-technology start-ups. The form of finance they provide, equity financing, provides a lump sum of capital up front in exchange for an ownership stake in the company. This eliminates the burden of monthly payments, for which most start-ups do not have the cash flow. Venture capitalists provide young companies with important assistance that helps them get up and running quickly. In this respect, they differ from banks and traditional financial institutions, which make "passive" investments and do not seek to participate in management activities. In the words of David Arscott, a Silicon Valley venture capitalist: "Venture capitalists provide a sounding board and a certain regimen for business development which helps turn good ideas into technical realities."[6] Venture capitalists frequently assist in recruiting the topflight personnel necessary to round out an entrepreneurial team. This credibility is perhaps most important when the company is trying to recruit a senior manager from a Fortune 500 company because such an individual often desires the reassurance regarding the company's future that the venture capitalist can provide. Of course, venture capitalists help companies raise additional rounds of financing by forming coinvestment syndicates with other venture capitalists.

Beyond this, venture capitalists confer a "seal of approval" on new companies that helps them establish working relationships with suppliers, vendors, banks, and customers. Through the use of their networks, venture capitalists help young companies secure legal counsel, accountants, public relations consultants, and a variety of ancillary business services. The assistance they provide lowers the obstacles facing new companies and enables them to produce new innovations and turn them into products as quickly as possible.

Venture capitalists keep the pressure on young companies to ensure that they develop products and grow as quickly as possible. Consequently, tensions can and often do arise between venture capitalists and the entrepreneurial group. Venture capitalists are judged and rewarded by maximizing the income of their limited partners over the ten-year life of their investment—not by the numbers of companies they launch or the number of jobs they create. Their goal is to build a company to the point where they can exit from and liquidate their investment with as much capital gains as possible, either by taking the company public or by merging it with another company. If a company is moving along too slowly or experiences business problems, venture capitalists can use their control over the board of directors or leverage over future inflows of capital to influence management. In more serious cases, they may replace top managers to protect their investment. Frequently, they may bring in a seasoned executive, typically one who is a known commodity who has previously been through the start-up process before and can supply motivation and management savvy needed to deliver. The founder is not necessarily fired but frequently receives a lower-level operating post such as R&D director, a nonoperating position on the board of directors, or a face-saving promotion. As one venture capitalist put it: "The venture capitalists are willing to be very kind and will promote the entrepreneur to chairman of the board as long as he remains outside the chain of command."[7] The venture capitalist's active involvement typically ends when the company goes public or is sold, although some stay on as members of the corporate board of directors.

A New Form of Finance Capital

What factors have shaped the rise of this new form of finance capital?

Even before the rise of industrial capitalism, there were informal sources of venture capital that came from independent financiers or "angels" whose vocation was to invest in high-risk companies founded by independent inventors. For example, expansion of the railroads was supported by a tightly knit group of "Yankee financiers" that was independent of major banks.[8] In his seminal writings on innovation in modern capitalism, Joseph Schumpeter emphasized the role of such independent financiers in supporting the new entrepreneurs of the industrial revolution.[9]

It was only in the post–World War II period that formal venture capital institutions began to emerge.[10] The earliest of these were tied to wealthy families such as the Rockefellers, Whitneys, Phipps, and Paysons.[11] In 1957, for example, Laurence Rockefeller's early venture fund, Rockefeller Bros., provided approximately $875,000 in capital for Itek, a new start-up.[12] Rockefeller's small family operation eventually grew into Venrock, one of the nation's leading venture capital funds. Another important venture fund, Bessemer Securities, emerged from the earlier activities of the Phipps family, while J. H. Whitney and Company became the investment vehicle for the Whitney family. Most of these early family venture capital funds grew up in New York City, the financial capital of the U.S. economy.[13]

In 1946 an important event took place. Wealthy Boston bankers and industrialists undertook to establish a formal venture capital vehicle tailored to the needs of high-technology business. The result of their effort was the creation of American Research and Development (ARD), the nation's first institutional venture capital fund.[14] They selected Georges Doriot, a professor of business at Harvard and a former army general, to head their new institution.[15]

Contemporaneously, a group of powerful bankers and industrialists began to sketch out plans for a new government agency devoted to the problems of small business and small business finance in particular. To lend legitimacy to their designs, they organized their efforts through the influential Committee for Economic Development, a group concerned with planning the postwar reconversion effort. The immediate result was the creation of the Small Business Administration in 1952.[16] By the close of the decade the federal government had established a new set of venture finance institutions under the auspices of the Small Business Administration known simply as "small business investment companies" or SBICs.[17] Under the SBIC program, the federal government provided significant subsidies to capitalize a new class of financial institutions catering to the demands of small business.[18]

The SBIC program created a new vehicle by which banks and other institutional investors could become involved in venture capital. During the early 1960s a large number of banks set up SBIC affiliates. In Boston the First National Bank of Boston established an SBIC and a pioneering program of loans for high-technology businesses. In San Francisco the Bank of America established its own SBIC affiliate, Small Business Enterprises. In Chicago two large banks, First Chicago and Continental Illinois, and a major insurance company, Allstate Insurance, launched major venture capital operations.[19] But New York City quickly emerged as the venture capital center for banks and financial institutions, as well as for family-based operations. During the 1950s, the 1960s, and most of the 1970s it consolidated its position as the nation's leading center for venture capital.

These early venture funds based in banks and financial institutions and the SBICs were the spawning ground for venture capital talent. In 1969 the head of Allstate's venture division, Ned Heizer, launched his own fund, Heizer Corporation, the largest independent fund of its day. On the East Coast, Peter Brooke, director of the Bank of Boston's venture loan program, founded TA Associates, currently one of the largest megafunds in the country, with assets exceeding $1.5 billion. And on the West Coast, George Quist, who ran Bank of America's SBIC, joined William Hambrecht to establish Hambrecht and Quist, a stock brokerage firm that developed very extensive ties to the venture capital community.

An interesting geographic pattern of venture finance emerged in the formative years. While the main sources of venture capital supply remained in old financial centers like New York and Chicago, the major foci for venture capital investments were the new high-technology areas like Silicon Valley and, to a lesser extent, the Route 128 area outside Boston. As mentioned previously, when the pioneering Silicon Valley start-up Fairchild Semiconductor was formed in 1957, its founding group turned for assistance to New York investment banker Arthur Rock. He in turn lined up Fairchild Camera, a corporation headquartered in Connecticut. Venture capital in the early years came to be characterized by a regular and predictable flow, from major financial centers like New York and Chicago to the burgeoning new outposts of high technology.

The Venture Capital Limited Partnership

By far the most critical factor in the emergence of the modern venture capital industry was the rise during the 1960s and 1970s of a new form of venture capital institution, the venture capital limited partnership. This new form opened up a whole new source of money for venture capital. Venture capitalists could now raise money from outside "limited partners"—banks, corporations, pension funds, and wealthy families. This capital would be controlled and invested by the professional venture capitalists, who function as "general partners."

The venture capital limited partnership was mainly a product of the emerging technology centers of Silicon Valley and Route 128. It enabled professional venture capitalists, who were often successful entrepreneurs, to establish new venture capital funds right in the heart of these booming high-technology regions. Many of the earliest, or model, limited partnerships were formed in Silicon Valley during the late 1950s and the 1960s.[20] Established in the late 1950s, Draper, Gaither and Anderson was the original Silicon Valley limited partnership. Many believe it was the first venture capital limited partnership in the nation. Draper, Gaither and Anderson itself was not successful, but it later sold its investment portfolio to Sutter Hill, a company that quickly emerged as one of Silicon Valley's leading venture funds. Another very early group of proto–venture capitalists revolved around an investment network of four or five informal investors known as "The Group," from which several subsequent partnerships later emerged.[21] A few early Silicon Valley venture funds were also initially constituted as SBICs. Among these were Frank Chambers's Continental Capital Corporation and Draper and Johnson, formed by William Draper, Jr., and Franklin "Pitch" Johnson, who would later launch an important limited partnership, Asset Management Corp.[22] During this early, formative period Silicon Valley venture capitalists began investing together in new companies and also started to share information on hot deals. By the mid-1960s these partnerships and funds had evolved into a network of sorts, the core of the new Western Association of Venture Capitalists.[23]

Of tremendous importance was the formation of the now legendary venture capital partnership of Davis and Rock. This Rock was none other than Arthur Rock, the investment banker who had arranged financing for Fairchild Semiconductor. His partner, Thomas Davis, was a Harvard-educated lawyer living in the Bay Area who worked in real estate development but invested a large amount of his time and money in high-technology start-ups. Though short-lived, the partnership of Davis and Rock was enormously successful (returning some $80 million on its original $3.2 million capitalization), making it a "model" that later venture capitalists would emulate.[24]

During the mid-1970s the early model charted by Davis and Rock was solidified with the formation of three major Silicon Valley partnerships: the Mayfield Fund launched by Davis himself; Kleiner Perkins, a fund formed by Fairchild cofounder Eugene Kleiner and Thomas Perkins, a high-ranking Hewlett-Packard executive; and Institutional Venture Associates, a fund launched by Reid Dennis, an executive with American Express and a member of The Group, and Burton McMurtry, a young venture capitalist with a proven track record. Like Davis and Rock before them, Dennis and McMurtry later split up to form their own independent funds, a pattern that would be repeated time and time again in the history of Silicon Valley venture capital. An important architect of this new model was the Silicon Valley law firm of Wilson, Sonsini, Goodrich, and Rosati, which helped draw up many of these early limited partnerships. The three venture capital partnerships proved the viability of the model both by attracting what was then considered to be a large volume of funds from institutional investors and by making a series of enormously successful investments in Silicon Valley start-ups.

By 1989 what had started in the 1960s as a dozen or so small-scale individual efforts had blossomed into a complex of more than two hundred venture capital partnerships with total assets in excess of $8 billion.

Boston, too, emerged as an early center for venture capital limited partnerships. During the 1960s and 1970s both ARD and "the bank" spawned a host of important limited partnerships. In 1963, for example, ARD alumnus Joseph Powell founded Boston Capital Corp. Other ARD alumni went on to launch important limited partnerships, such as Palmer, Greylock, and Morgan Holland, and other leading Boston venture capital funds. In 1968 Peter Brooke launched TA Associates, which begat a host of other venture capital funds, such as Burr, Egan and Deleage and Claflan Capital Management.[25]

The "Gold Rush" Years

The rise of the limited partnership helped fuel the massive venture capital boom of the 1980s. Because it enabled large financial institutions to invest in venture capital funds, the limited partnership dramatically increased the potential pool of venture capital.

Between 1980 and 1989, the volume of venture capital increased more than tenfold, skyrocketing from less than $3.5 billion to more than $33 billion (see figure 4.2).[26] A variety of factors contributed to this venture capital "gold rush." First and foremost was the extraordinarily high rate of return generated by investments in high-technology start-ups and in venture capital funds. Starting from small beginnings in the 1960s and 1970s, venture capitalists had scored tremendous successes with investments in electronics start-ups such as Fairchild, DEC, and Intel. By the late-1970s and 1980s it was clear that venture capital investments produced huge returns. According to one estimate, the returns from venture capital funds were roughly five times greater than returns

Figure 4.2   The Venture Capital Explosion of the 1970s and 1980s

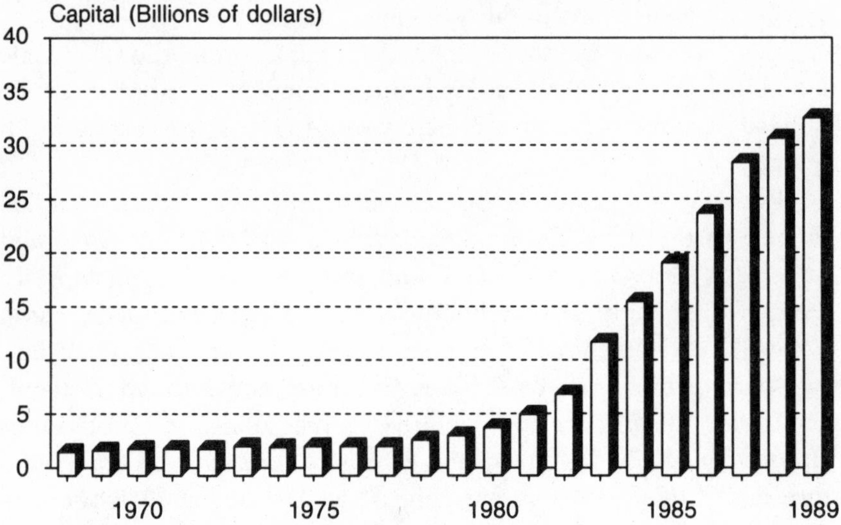

Capital (Billions of dollars)

SOURCE: *Venture Capital Journal,* various issues.
NOTE: Figures for 1969–76 are based on an estimated nominal pool of $2.5 billion. 1989 data are from Venture Economics Estimates.

on corporate stocks and bonds (see figure 4.3).[27] This exerted a powerful pull on outside capital, which poured into venture capital. Success itself was reinforcing: early successes enriched the coffers of venture capitalists, who plowed that money right back into new investments. Venture capitalists could now invest even more in start-ups, increasing their odds of success. And they could back entirely new technological opportunities, like biotechnology, which were certain to take far longer than electronics to actually develop and ship products. Success had a powerful demonstration effect, encouraging more entrepreneurs to form companies and opening up additional investment opportunities for venture capitalists.

The explosion in venture capital was also accelerated by sharp declines in the overall rate of profit, which prompted an investment shift from basic industry toward new industries and more speculative investment opportunities like venture capital.[28] As is now well known, the U.S. manufacturing industry began to experience a multidimensional crisis of profits and productivity during the mid-1960s, a crisis that continued through the 1970s and 1980s.[29] Inflation became a

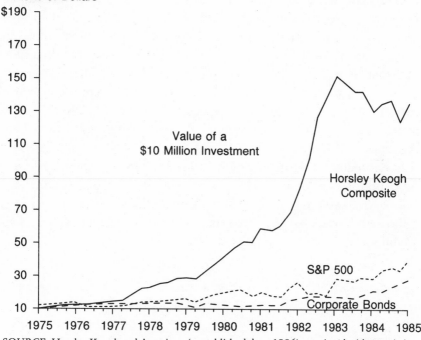

Figure 4.3   Venture Capital Returns Compared with Returns on Corporate Stocks and Bonds (time weighted returns over ten years)

SOURCE: Horsley Keogh and Associates (unpublished data, 1986); reprinted with permission.

perennial feature of the economic scene, as large corporations—trapped by old institutional arrangements—responded to pressure for wage increases by passing them on to consumers and thereby setting off a tenacious wage-price spiral. The domestic market for consumer durables became saturated, foreign trade slipped, and the U.S. corporations' share of world markets for manufacturing declined substantially. As a result, corporations, banks, and other financial institutions slowed investments in production and R&D, and some of that capital made its way into other outlets. And a portion of this capital spilled over into venture capital–related activities.

Government also played a role by reducing the capital gains tax rate and loosening restrictions governing pension fund investments. These two public policy measures increased the attractiveness of venture capital and led a variety of institutions to increase their venture capital investments.

The gold rush years saw Silicon Valley and Route 128 consolidate their lead as the main centers for venture capital investment. In 1988 Silicon Valley attracted more than $1.2 billion in venture capital investment, 40 percent of the national total. The Route 128 area attracted one-quarter of this, $330 million dollars. New York attracted just $150 million in venture capital investment, less that one-half that of Boston and roughly one-tenth that of Silicon Valley.[30] These years saw an even more dramatic shift in the center of venture capital supply, from older financial centers of New York and Chicago to the new high-technology regions of Silicon Valley and Route 128. As figure 4.4 shows, by 1988 California had become the nation's leading center of venture capital, with $8.1 billion compared with $7.6 billion for New York. Boston was in third place with $4.8 billion, followed by Connecticut with $1.8 billion, Chicago with $1.6 billion, and Texas with $1.1 billion. No other state claimed more than $1 billion in venture capital.[31]

Figure 4.4 Major Centers of Venture Capital in the United States (figures in millions of dollars)

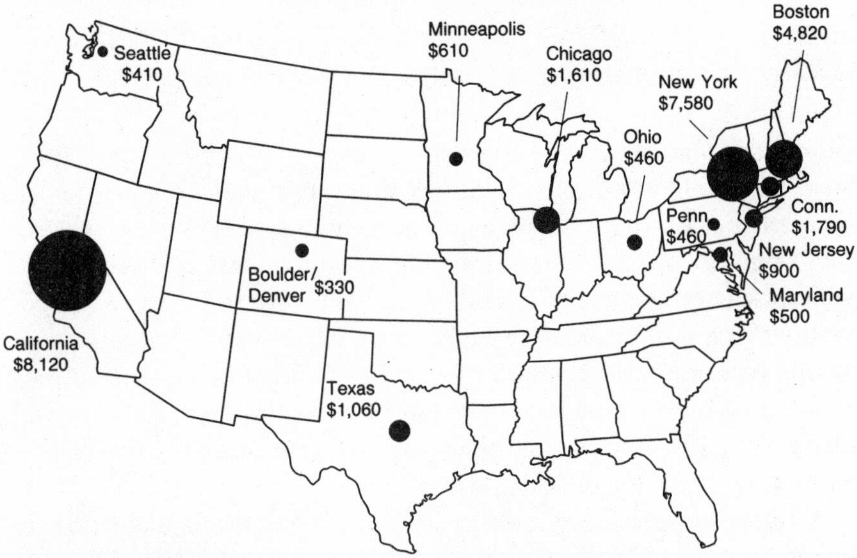

SOURCE: *Venture Economics* (January 1990); reprinted with permission.
NOTE: Complexes shown have 1 percent or more of the national venture capital pool of $31,100 million.

68

High-Tech Capital

There are important differences between the new venture capital centers of Silicon Valley and Route 128 and the older financial centers of venture capital, New York and Chicago.[32]

Silicon Valley and Route 128 venture capitalists invest mainly in local companies, whereas financial venture capitalists in New York and Chicago do not. According to analyses we have conducted of a large database of venture capital investments, Silicon Valley venture capitalists invest nearly three-quarters of their venture capital in local companies, and Boston venture capitalists invest roughly half of their capital locally. This stands in sharp contrast to venture capitalists in financial centers like New York and Chicago, who export the great bulk of their funds, mainly to Silicon Valley and Route 128. According to our own estimates, New York City venture capitalists make nearly half (43 percent) of their investments in Silicon Valley companies, and just 6 percent of their investments in New York companies.[33]

Silicon Valley and Route 128 venture capitalists are hands-on investors. They are located close to their investments and are heavily involved in the important strategic and managerial decisions of the companies they back. Financial venture capitalists in New York and Chicago are located much farther from their investments and often take a hands-off approach, depending instead upon a local "lead" investor to monitor the situation. Because of this, a symbiotic relationship has grown up between the two groups of venture capitalists: venture capitalists in high-technology regions make and manage investments and look to their counterparts in financial centers as a source of additional capital. According to our estimates, New York venture capitalists make twice as many (36 percent) of their investments with California venture capitalists as they do with other New York venture capitalists (18 percent).[34]

Venture capitalists in Silicon Valley and the Route 128 area are linked in dense communities, or networks. In both areas, venture capitalists meet frequently to discuss mutual investments and potential deals, and they often locate their offices close to one another to promote contact and interaction. For example, a large proportion of Silicon Valley's leading funds are located along just two roads, Page Mill Road in Palo Alto and Sand Hill Road in Menlo Park. Literally dozens of Silicon Valley venture capitalists have their offices in a single office complex at 3000 Sand Hill Road. They are able to congregate in each other's offices and at the restaurant in the center of the complex to discuss and develop mutual investments. Such tightly spun networks enable venture capital networks to locate better investments, pool funds, share risk, and learn from one another's successes and failures. Thus, California venture capitalists make nearly half (44 percent) of their investments in syndicates with other California venture capitalists.[35]

Most significantly, the venture capital networks of Silicon Valley and Route 128 are embedded in the broader fabric of local high technology. Venture capitalists, a component of the well-developed social structures of innovation that characterize these leading high-technology regions, bring important financial resources and business development skills to those networks. Interestingly, a growing number of New York venture capital funds have opened branch offices in Silicon Valley and the Route 128 area to gain better access to these networks. Citicorp, for example, has a large branch office in Silicon Valley, while Bessemer Securities has branch offices in both Silicon Valley and the Route 128 area.

However, in contrast to what some believe, venture capital alone cannot stimulate either high-technology development or breakthrough innovation.[36] It can only supplement and reinforce this process in places that already have well-developed social structures of innovation. Consider the remarks of Daniel Holland of Morgan Holland, one of Boston's leading venture capital funds: "It is not venture capital that is the start of entrepreneurial activity. You can't simply put six venture capitalists in Butte, Montana, and expect that the availability of venture capital will engender a Route 128."[37]

The tendency of venture capital to flow toward established technology centers like Silicon Valley and Route 128 raises serious questions about the efficacy of state and local policies that aim to stimulate high-technology development by increasing the local availability of venture capital. Since most of these areas lack the requisite technology base, such policies are likely to have perverse effects as local venture funds are either invested in unprofitable local investments or exported to established technology areas.

The Breakthrough Bias

What does this new form of finance capital mean to American high technology?

While venture capital quickens the overall pace of technological innovation, it biases the U.S. high-technology system in the direction of breakthrough innovations of the sort that open up new markets and produce huge returns on investment. The reasons for this are built into the very mechanisms of venture capital investing itself.

Venture capitalists as well as high-tech entrepreneurs bet on the superprofits that come from companies that create breakthrough products, open up new markets, and at times forge whole new industrial sectors. David Morgenthaler, Sr., former head of the National Association of Venture Capitalists and a prominent venture capitalist, told us that he looks for companies that can cross the $500 million sales hurdle in seven years.[38] And that is exactly what the great successes such as Apple, Compaq, or Sun Microsystems did.

Finding companies that can deliver such superprofits is obviously risky. Venture capital is a form of finance capital that has evolved to cope with these odds. In contrast to most other forms of finance capital, venture capital is organized to operate with success rates of one in ten, or even one in twenty, and still generate excellent returns. The mathematics of the breakthrough economy is such that a relatively few big successes can easily offset a huge number of failures. One Apple more than offsets a dozen or so outright failures. The equity position is the reason that the venture capitalist is able to profit from extremely risky investments: The potential capital gain from success is limitless and more than compensates for failures. In the words of Andrew Rachleff, a Silicon Valley venture capitalist: "We're in the business of hitting home runs."[39] Indeed, the system is rigged so that home runs more than compensate for strikeouts.

Figure 4.5   Venture Capitalists Generate High Rate of Return with a
Few Big Winners

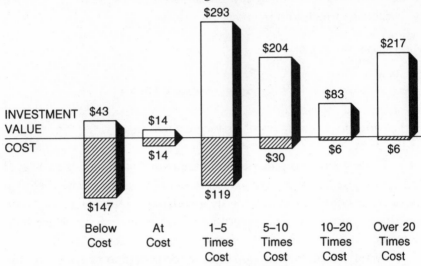

SOURCE: Horsley Keogh and Associates (unpublished data, 1986); reprinted with permission.

In this kind of system breakthroughs are not only a good bet, they are the best bet available. A study by Horsley Keogh and Associates evaluated the performance of more that five hundred portfolio investments made by ten leading venture capital partnerships.* According to their analysis, only about one in ten venture investments were successes. But the returns they generated were more than enough to cover the out-and-out failures and the stagnant companies that venture capitalists call the "living dead"—companies that make things, employ people, and generate some small amount of revenue but are valueless from their home run perspective. In fact, 13 percent of all investments accounted for $504 million, or 59 percent of $854 million in market value (see figure 4.5). More than half of all investments made by these funds went nowhere; they either broke even or lost money.

Venture capitalists operate under extremely stringent time pres-

*This type of data is very difficult to secure because venture capital partnerships are fully private entities and hence not subject to disclosure laws. But Horsley Keogh is a "fund of funds"; that is, it places investments in a number of venture capital partnerships, so it has access to data on the performance of various funds. Horsley Keogh and Associates, *1985 Horsley Keogh Venture Study* (San Francisco, Cal.: Horsley Keogh and Associates, 1986).

sures and they transmit this to their clients. As we have seen, they need to enter and exit investments in five to seven years so they can recoup their investments, ensure adequate returns to limited partners, and establish the kind of track record needed to launch future funds. It is astounding to think that the five- to seven-year time horizons of venture capitalists are considered "long-term." But, given the current business environment in which quarterly earnings are the rule, this may indeed be as long-term as it gets. This means that there is tremendous pressure to back potential home runs and to pull the plug on weak starters. A self-fulfilling prophecy can result: companies that do not perform up to some threshold level are quickly scuttled, while others are built up. In the short run, this may unnecessarily sacrifice some companies that have the seeds of important innovations or, at least, the potential for steady growth. In the longer run, it may seriously dilute the overall high-technology base, especially if the losers are concentrated in particular technology sectors. In the words of a Silicon Valley venture capitalist: "If you have too much of a short-term focus, it will discourage you from doing important investments in companies with core technologies."[40]

Other companies may be "ramped up" for an initial public offering well before they are ready. Venture capitalists will put incredible pressure on a company they believe has a chance of success. Employees may be forced to work around the clock to cross technological hurdles and meet product deadlines, especially since venture capitalists use such milestones to justify further investments and an increase in the company's stock valuation. Public relations firms will create the expectation of great things even when there is little reality to these claims. Many biotechnology companies, for example, went public before they had products or even a steady cash flow. Not surprisingly, it is very difficult for fledgling companies to become established, self-reliant enterprises in this pressure cooker environment.

And once a company goes public, it runs a great risk of being abandoned. Since most of the key actors have already made their killings, they have every incentive to leave and start the whole process over again. But the companies they have built still have to deliver products and make a profit. Abandonment by just a few key people can reduce a once-promising company to an empty shell. This pattern of "throwaway" entrepreneurship squanders scarce resources and costs U.S. high technology dearly in terms its international competitiveness in the long run. Beyond this, the pressure to generate commercial breakthroughs may cause some venture capitalists to orchestrate raids of established high-technology firms (see chapter 5) or finance "copycat" companies (see chapter 6), with catastrophic effect on the technological competitiveness of the United States. Gordon Moore of Intel is quite explicit: "The possibility exists of there being too much venture capital, in which case everything gets fragmented and all continuity in things that established companies do gets lost. . . . Right now there are venture capitalists chasing awfully hard to get money invested. I believe they are supporting a lot of things that don't deserve support."[41]

Our system of venture capital–financed innovation ultimately exerts an overwhelming "selection bias" in favor of companies and products that can produce big breakthroughs, crowding out other forms of innovative activity. In this sense, venture capitalists act as technological gatekeepers for the breakthrough economy, helping to determine which new technologies are admitted. While venture capital is good at generating breakthroughs, it is not geared to developing the other types of innovations that contribute to follow-through.

Venture capitalists neglect important, though less profitable, product refinement innovations and improvements in manufacturing processes. The reason for this is easy enough to understand. Venture capital financing does not provide enough capital to undertake big improvements in manufacturing processes or to refine and improve products continually. Moreover, process innovations seldom offer the kinds of returns venture capitalists are looking for, except when they are linked to breakthrough products. So, for example, venture capitalists have invested in biotechnology in the search for new breakthrough products, neglecting important process innovations in this field. This emphasis has channeled U.S. research into areas in which discrete breakthroughs can be made, such as new pharmaceutical products, and away from industrial process applications like beer fermentation or the production of organic chemicals. And venture capitalists also ignore long-term basic R&D in favor of commercial breakthroughs with large potential markets. According to John Wilson, a founder of one of Silicon Valley's most prestigious high-technology law firms, Wilson, Sonsini, Goodrich, and Rosati: "Venture capitalists tend to want to take things that can be done very rapidly. Long-term R&D does not get done."[42]

The tremendous financial rewards associated with venture capital–financed innovation exert a powerful pull on highly skilled researchers and technologists, drawing them away from industrial and manufacturing engineering. Since the pool of high-technology talent is relatively fixed, this ".attraction effect" creates a situation in which a large and growing share of our technical base concentrates on breakthroughs, leaving a dwindling share to worry about follow-through. Venture capital–financed innovation likewise pulls university researchers away from long-term basic research toward more applied areas and can cause important conflict of interest problems.[43] This is especially evident in the large number of university spin-offs that have emerged in biotechnology and superconductors, and also in a more general climate that favors university researchers engaging in joint R&D with industry.

This all-encompassing quest for superreturns may actually weaken American high technology in the long run. While venture capital may be the best way of financing a high-technology sprint, the unfortunate reality is that the quest to secure the benefits of high-technology competition increasingly resembles a marathon.

II

.

LIMITS

5

The Hypermobility of High-Tech Labor

∎

In Silicon Valley, if somebody wants to change jobs, all they have to do is turn into a different parking lot off a different freeway exit.

—JERRY SANDERS, CEO, Advanced Micro Devices[1]

In Silicon Valley, personal loyalty replaces corporate loyalty. This scrapping for people will get much worse.

—JAMES KOFORD, director of research, LSI Logic[2]

The ability of high-technology workers to move around, change jobs, and form new businesses has typically been hailed as the great strength of our high-technology system. While many continue to praise this new era of entrepreneurship,[3] a growing number of dissenters, including a number of top high-technology executives, have recently begun to question the efficacy of our system of unconstrained entrepreneurship and labor mobility. Andrew Grove of Intel, for example, has publicly attacked the gold rush mentality of start-ups, and Gordon Moore also of Intel has openly criticized what he calls "vulture capitalists" for orchestrating raids of top R&D scientists and high-technology managers. Other entrepreneurs, such as Jerry Sanders of Advanced Micro Devices (AMD) and Charles Sporck of National Semiconductor, have similarly questioned this system of "free and unfettered" labor mobility. They suggest that high levels of turnover and defection disrupt the continuity of our high-technology effort and weaken our ability to compete in global high technology. *Electronic Business,* a leading trade journal of high technology electronics, neatly sums up this new reality with the phrase "the case of the walking brain."[4]

These problematic features of the breakthrough economy—the high levels of turnover, defections, and raids—can be traced to a deeper phenomenon that we call the "hypermobility of high-tech labor."* This hypermobility is rooted in the new organizations and incentives of the breakthrough economy, which enable high-tech think-workers to secure a share of the superprofits they generate. The system operates to enable them to take multiple shots at generating and gaining superprofits. High-technology labor markets provide a wealth of job opportunities for think-workers, making it easy for them to move from job to job, start-up to start-up—and to do so with little risk.

But what's good for individuals may not be good for American high technology. The hypermobility of high-tech labor is aimed at short-term returns rather than long-term performance. And increasingly, actions that ensure short-term financial gain are different from those necessary to build successful high-technology firms and industries. Hypermobility results in disrupted research teams, exorbitant salaries, wasted resources, burned-out workers, and a redistribution of technical talent from follow-through activities to breakthrough innovation.

Most significantly, hypermobility has made it increasingly difficult for established companies to internalize cycles of innovations. Instead, we face a situation in which each new innovation results in the creation of new companies, because firms are in effect unable to contain the "product technology cycle" internally. Rather than building stable competitive companies, we develop one-shot, breakthrough firms. Hypermobility has become a fundamental dilemma of the breakthrough economy—one to which there is no easy resolution.

---

*This form of hypermobility should not be confused with the high turnover rates experienced by high-technology manual laborers. The former is usually voluntary; the latter is often not. This chapter focuses on the hypermobility of high-tech mental labor only.

Turnover and Defection

High-technology companies in Silicon Valley and the Route 128 area face turnover rates that are truly astounding—25 to 35 percent or even 50 percent per year.[5] These extraordinarily high rates have helped to keep the overall turnover rate for the electronics industry quite high in recent years: 17 percent in 1984, 13.5 percent in 1985, and 23 percent in 1986.[6] Leading analysts predict that turnover will increase to more than 30 percent by the mid-1990s.[7] Only a few large high-technology companies, like Intel, Hewlett-Packard, and DEC, have more manageable turnover rates of about 10 percent. According to the venture capitalist Arthur Rock, the turnover rates experienced by many companies are so high that "the amount of time an engineer spends at a new company is less than the time it takes to develop a new product."[8] In a recent survey, CEOs of leading high-technology electronics companies named "attracting and keeping key people" as one of the biggest challenges of the 1990s.[9]

While accurate data on turnover and mobility for high-technology start-ups are hard to secure, a study published in 1989 provides important new evidence of the extent of hypermobility.[10] This study tracked the career and mobility patterns of 275 semiconductor production engineers between the years 1980 and 1986. Over this period more than three-quarters of the engineers changed jobs at least once, with the annual rate of new job "starts" averaging between twenty-five and thirty-five per hundred and the annual rate of job "separations" from quits and layoffs averaging between twelve and twenty per hundred.

As might be expected, mobility was greatest among Silicon Valley engineers, who accounted for roughly half of all employment changes. And most of the job changes among Silicon Valley engineers took place between employers located within just a few miles of one another. More than three-quarters of the engineers who quit jobs in Silicon Valley took new jobs with other local companies. In contrast, engineers from semiconductor firms located elsewhere, for example, Texas Instruments and Motorola, were much more likely stay in one job for longer periods. Based on these findings, the study concludes that "interfirm mobility is a persistent feature of the labor market for semiconductor engineers, rather than an occasional symptom of job shopping."[11]

High rates of mobility are also prevalent in the biotechnology industry, but there it takes a different form. The most common avenue for mobility in the biotechnology industry is job-hopping among established firms—both existing start-ups and large chemical and pharmaceutical companies. In 1983, for example, Cetus raided Bio-Rad Laboratories and secured the services of the researcher and marketing analyst who were the heart of Bio-Rad's cancer diagnostic program. Cetus also lured Biogen's president, Robert Fildes. The biotechnology industry has not seen a high level of departures to form new start-ups, however. One reason for this is that biotechnology is closely tied to basic science, so university professors are commonly the main source of new start-ups. A second reason is that biotechnology products take a long time to develop, especially medical and agricultural products, which are subject to lengthy government reviews. Thus, new biotechnology start-ups are more likely to be launched by university professors operating in wholly new application areas than by spinoffs from existing companies.[12]

High-technology professionals leave high-paying jobs at innovative companies for a number of reasons. Some go to better positions with competitors. Others want to join new start-ups. Still others launch their own company. Some leave of their own volition; but many are lured away. Most frequently, the goal is not more salary but an equity stake, because that offers the opportunity to become very rich. Wilfred Corrigan, former Fairchild CEO and current CEO of LSI Logic, gives the following answer to the question: Why do think-workers defect? "The answer is simple. Equity and challenge. The bright people in this business are always looking for a big chance, for something new. People like being superstars, IQ likes being with IQ. That gets diluted in a big company."[13]

Think-workers are motivated by the excitement of the highly innovative, breakthrough environment afforded by a new start-up. They want to be on an exciting project team or in a company that is "going places," even if it means leaving a secure job for a risky proposition. Of course, the provision of large equity stakes can make even the riskiest propositions look attractive.

Defections tend to increase after major projects have been completed, when members of successful R&D groups often suffer from "postpartum" depression. Tracy Kidder's *The Soul of a New Machine* chronicles the dispersal of Data General's Eagle computer project team. Members of the development team became dissatisfied as soon as the project was completed. They began to resent management for breaking up the team and for insufficiently recognizing their achievements. He quotes a team member: "It was a group that was formed and achieved this remarkable thing, and the company has deemed to reward the group by blowing it up. It's really sad."[14]

The point here is a basic one. Once a project has been completed and a product developed, R&D scientists and engineers find it exceedingly difficult to return to mundane activities and routines. Mobility is a way to continually re-create that exciting, challenging environment—and be rewarded financially to boot. Defection can be a self-reinforcing process as well. Departures of one or two critical employees can create an "emulation effect," causing many more to leave. Founders usually recruit people from their previous employer, and start-ups may draw significant numbers of people from as few as one or two established firms.

In one sense, hypermobility is a "good thing" for high-technology professionals. Their ability to leave forces companies to take their demands seriously. Companies go to great lengths to contain defections by offering workers higher salaries, increased equity or ownership stakes, better work environments, and more discretion over their work.

Raids and "Vulture Capitalism"

Raids, unlike defections, are initiated by actors outside the company, whether established firms, new start-up companies, or venture capitalists. Robert Noyce, who disapproves of raids, described the difference between these and defections as follows: "I don't think it is legitimate for people to reach into existing companies where there is a technology they want to exploit and drag it out. If the individual is unhappy and wants to leave, that's one thing. But to have outside influences come in and create dissatisfactions feels uncomfortable to me."[15] Venture capitalists and high-technology executives speak of a somewhat fuzzy "trigger point," when a series of hirings from a single company becomes a raid. The movement of two or three employees to a new start-up is seen as fair game; the movement of more in a short period of time can be cause for legal action. James Koford of LSI Logic put it this way: "Mobility in groups of three or even four is OK. The problem comes with really big, well-financed start-ups. It's raiding when they get too many people too close together from one company. That does happen."[16]

It is difficult for a company to defend itself against raids because the raider can drastically increase an individual's salary and provide a bigger equity stake. The company being raided may find it impossible to match the new offer without upsetting its existing remuneration structure. Larry Jordan of Seeq explains: "It's impossible to counter certain types of raids, particularly those by start-ups. If they need a certain kind of skill, they might double someone's wages or salary to get it. We can't meet the offer because we might have fifteen other people at the same level whom we'd have to pay more as well. So it becomes an impossible situation."[17]

In recent years many entrepreneurs and even some venture capitalists have begun to think that the increased proliferation of start-ups is due to what they term "vulture capitalism," as venture capitalists actively lure employees away from their jobs at established companies. They complain that such actions worsen the problems of defection and turnover, a situation that makes it hard for high-technology companies to retain people, develop products, and compete on a global scale. In some cases venture capitalists have actually orchestrated raids, going to extreme lengths to entice key managers or researchers from established companies. As chapter 4 shows, the bottom-line objective of venture capitalists is to make money. If a particular individual or group is needed to make a company successful, chances are venture capitalists will assist in efforts to attract that person.

Still it is quite difficult to say what is and what is not vulture capitalism. Take the following case.[18] In the early 1980s an executive within a well-known Silicon Valley high-tech company received a phone call from a venture capitalist requesting information on a group of former subordinates who were planning to start a new company. During the conversation it became apparent to the venture capitalist that the executive might be the right person to launch a new company in this new, potentially lucrative field. Within a week, the executive and the venture capitalist met on the East Coast to discuss the possibility of launching a start-up. The next day the executive met with the senior partner of another well-known venture capital fund. Within a month the executive was writing a business plan and beginning to recruit personnel from his former employer. While the new company zoomed into the limelight, the old employer lost a bevy of key researchers, top managers, and important marketing executives. Even though venture capital had catalyzed an important new start-up, it had imposed tremendous losses on the existing company.

An informal set of rules, or norms, has developed to "regulate" recruitment behavior or raids in areas such as Silicon Valley and Route 128. For example, companies that are experiencing hard times or where morale is low are seen as legitimate targets for raiding. Venture capitalists and entrepreneurs who engage in other types of raids are considered to be disreputable and may face informal sanction from their peers. All the venture capitalists we interviewed said that they refrained from overt raids, because it can hurt their reputation and future ability to attract talented entrepreneurs and management talent.[19] That said, informal sanctions are by nature weak and unenforceable. Like beauty in the eye of the beholder, it is often very difficult to tell the difference between a raid and smart business behavior. And successful companies frequently become immune to criticism, confirming the old adage "success covers all sins."

Taken in isolation, each raid, each example of vulture capitalism, may not seem all that damaging. And of course, for the supporters of our entrepreneurial high-technology system, any new business formation is cause for celebration. But in fact the current pervasive pattern of raids and defections threatens to weaken seriously our ability to develop and implement high technology.

Chronic Entrepreneurship

The phenomenon of hypermobility underlies what Robert Reich refers to as "chronic entrepreneurship," which stems from employee defections to form new start-ups.[20] According to the most recent data available from the U.S. Small Business Administration, more than 100,000 new high-technology companies were launched during the decade stretching from 1976 to 1986 (roughly 10,000 per year), more than 60,000 or 6,000 per year of which were small start-ups employing under twenty people. According to the most recent data available from *Venture Economics,* venture capitalists backed 1,338 new companies in 1988.[21]

An Intel executive provides striking insight into the process behind these numbers, drawing a compelling comparison with the Broadway theater industry. "Most [Silicon Valley] companies are like so many Broadway plays. The venture capitalist is like the producer. An itinerant group of 'actors' get cast in the needed roles. The 'play' opens—has a 'run' (short or long)—then it closes. Time to put a new play together."[22] The effect of this mass start-up activity is constantly to drain huge numbers of people away from established high-technology firms, seriously weakening their ability to compete against Japan's integrated corporate giants.

Intel is often pointed to as a victim of chronic entrepreneurship, and indeed past and current Intel officials have been among the most vocal critics of this proliferation of new spin-offs and start-ups. Intel has in truth been burned by a wave of recent spin-offs. Roughly fifty Intel engineers defected to Daisy Systems, a pioneering workstation company. Most of MIPS's microprocessor engineers were hired away from Intel. Zilog was formed by seven former Intel employees, and Sequent Computer System by eleven defecting Intel engineers. Seeq was launched by two senior managers from Intel's Special Products Division.[23] And one of Seeq's founders, Gordon Campbell, later left Seeq to form Chips and Technologies, a company producing Intel-like components for IBM personal computer clones.[24]

For some, the self-righteousness of Intel's top management may well seem self-serving. As the venture capitalist Donald Valentine succinctly put it: "My friends at Intel just don't want anyone do what they did. Was it right in 1969 or 1970? Is it right now? To claim national interest around the corporate flag is bullshit."[25]

In any case, chronic entrepreneurship is not a new phenomenon. It is virtually built into the breakthrough model of high-technology organization. Indeed, the roots of chronic entrepreneurship date back to the very beginnings of entrepreneurial high technology, to the birth of Fairchild, which itself was formed as a result of a mass defection (see chapter 3). Indeed, Fairchild's evolution provides a telling case study, a microcosm of sorts, for illustrating the dynamics and the long-run costs of chronic entrepreneurship.[26]

Large-scale defections from Fairchild began right after it was founded. Six employees defected to Rheem Semiconductor in 1959, less than two years after Fairchild was launched. Amelco was launched in 1961 by three of the original Fairchild founders. Four more Fairchild employees formed Signetics in 1961, another started Molectro in 1962, and two more launched General Microelectronics in 1963.[27] When General Microelectronics was acquired by Philco-Ford in the late 1960s, a number of its employees left to form another company, American Microsystems, initiating the second generation of Fairchild spin-offs.[28] By 1963, barely six years after the company was founded, only four of the eight original founders of Fairchild remained with the company. Some of the defections were so serious that Fairchild brought lawsuits in an unsuccessful effort to stanch the outflow of its technology. In one case, Rheem paid Fairchild some $75,000 in a court settlement. In another, General Microelectronics was forced to retreat into an area that was not directly competitive with Fairchild. To move into this new area, General recruited Fairchild's leading expert in metal oxide semiconductors; he then made General the leader in that important emerging technology.

Throughout the late 1960s and early 1970s, Fairchild was buffeted by a series of mass defections that were much more costly than the spurt of recent defections from Intel. Intel itself was formed as a result of a mass defection of top Fairchild executives, including CEO Robert Noyce, Gordon Moore, and Andrew Grove.[29] Charles Sporck, another top Fairchild executive, left to become president of

National Semiconductor. He took with him a large group of Fairchild employees.[30] One industry insider describes the mass defection that followed Sporck: "Charlie emptied Fairchild of talent. He went over and took his best manufacturing guy, a fellow named Bob Mullen, who now runs some company in San Diego. He took Pierre Lamond, who is now with Capital Management, who was his second-best manufacturing guy. Then they took the next level, then the foremen."[31] In 1969 Jerry Sanders took a large group of employees to form American Micro Devices. A host of other less important companies—Four Phase, Qualidyne, Computer Micro Technology, Advanced Memory Systems, and Precision Monolithics—were also launched by Fairchild defectors. Once this nation's premier semiconductor company, Fairchild was badly weakened by these defections. The head of public relations at American Micro Devices observed: "Jerry [Sanders], Bob Noyce and Charlie Sporck were the three legs of the tripod upon which Fairchild rested. You had Bob Noyce inventing them, Charlie Sporck making them, and Jerry Sanders peddling them. And, they remain the three biggest names in our industry."[32]

Fairchild responded to the loss of its top management by raiding Motorola's Lester Hogan and his entire senior management team. Hogan himself was no stranger to this game. While at Motorola he had recruited a large group of engineers from GE. In return for his move to Fairchild, Hogan received a then unheard of starting salary of $120,000, 10,000 shares of Fairchild stock, and a $5 million loan for options on 90,000 more. According to some sources, Hogan's remuneration package was more than double what former president Robert Noyce had been getting. Hogan was replaced by one of his Motorola protégés, Wilfred Corrigan, who later left to start LSI Logic in the early 1980s.[33]

Fairchild never completely recovered from the defections of the late 1960s and the 1970s. After stumbling along under Hogan and Corrigan, the company was acquired by the Dutch oil drilling giant Schlumberger in 1979, only to be put up for sale again 1983. In a much-publicized case, the U.S. government prevented Japanese electronics giant Fujitsu from acquiring Fairchild in the mid-1980s. By 1987 a weakened Fairchild was bought by National Semiconductor and merged out of existence.[34] The slow, steady decline of Fairchild, which was once the premier semiconductor company in the world, stands witness to the staggering cost of chronic entrepreneurship.

Chronic entrepreneurship and mass defection are also prevalent in the computer industry. Data General, for example, was formed by Digital Equipment Corporation's Edson DeCastro and two of DEC's top computer designers.[35] Cray Research, our leading supercomputer company, was founded when Seymour Cray left Control Data Corporation.[36] In 1987 one of Cray's leading computer designers, Steven Chen, left with forty-five other Cray employees to form a new supercomputer company funded in part by IBM.[37] Now Seymour Cray himself has left to form yet another company, Cray Computer in Colorado Springs, Colorado. Apollo, the first company to market workstations, was formed when William Poduska left Prime Computer. Poduska later left Apollo to form another Boston area start-up, Stellar Computer. Convex, a recent minisupercomputer start-up, was launched by two Data General computer designers. Alliant, another minisupercomputer start-up, was launched by another group of Data General employees.[38] Mentor Graphics, a leading design automation company, was founded by Thomas Bruggere, an engineering manager for Tektronix, nine other Tektronix employees, seven software engineers, a finance person, and a marketing executive.

While the constant drive to capture the superprofits of breakthrough innovation is a primary motivation, chronic entrepreneurship is exacerbated by the new organizational structures of the breakthrough economy. Thus, the independent R&D project teams (discussed in chapter 3) can be thought of as "potential start-ups." While independent or quasi-independent research teams can be a valuable source of new ideas, they can easily get too isolated, too cut off from the rest of the company. This is especially true of what people in the industry term "skunkworks," or teams of engineers and managers working on a project in an environment separated from the rest of the firm.[39]

Defections of development teams can be provoked in a number of ways. Disagreements between internal groups and top management can precipitate a spinoff, especially when funding cuts or cancellation of a pet project are involved. Management, for example, may not be willing or able to provide the resources that the team feels are needed for them to succeed. Internal esprit de corps can grow so intense that the team decides to go off on its own. And, of course, venture capitalists may lure team members or the entire team away with the promise of tremendous financial gain.

Hypermobility: The Underlying Cause

Defections and raids, chronic entrepreneurship, and vulture capitalism are all expressions of the hypermobility of high-tech labor, which is itself rooted in the organizational structures and economic incentives of the breakthrough economy. As we have seen, the breakthrough economy is based upon the superprofits that come from breakthrough innovation and high "capital gains potential." The provision of huge equity stakes is necessary to attract qualified think-workers, the human repositories of breakthrough innovations. Given this system, high-technology think-workers maximize their chances of accruing large capital gains by moving from start-up to start-up. In fact, employees are most likely to leave after their original stock option plans or "golden handcuffs" vest (usually five years) and try again with a new start-up. In this way, high technology think-workers play the odds by taking multiple shots at latching on to a successful start-up. This system provides a new twist on the age-old distinction between capitalist-owners and salaried workers: high-tech think-workers have multiple chances to transform themselves from salaried employees into company owners.

Changing jobs like this involves little risk because demand for R&D scientists, engineers, and top-notch marketing personnel and managers far outstrips supply. And as long as technology continues to change rapidly and markets continue to grow, there is little possibility of an equilibrium developing. It is easy for an engineer or manager to walk away from a job when a host of other opportunities is readily available.

One of the most interesting aspects of the labor market for high-technology think-workers is that they can create their own employment. They do not depend exclusively upon the external labor market, because they are able to form their own firms or work as independent consultants when they are between jobs. This creates its own paradoxes: an individual can go from being an employee to an owner of a company, and then, if the company fails, to being an employee once again. Hewlett-Packard actually has a policy of rehiring employees who left in good standing to do a start-up.[40]

As an individual strategy, then, hypermobility tends to increase a think-worker's value. Being a part of a start-up—or better yet being a part of a series of start-ups—provides the hands-on experience and the aura of success entrepreneurial companies and venture capitalists look for. Mobility enhances the development of personal networks, which are a critical source of inside information, technical assistance, and employment and investment opportunities. Hypermobility confers distinct benefits on those able to take advantage of it.

Costs of Hypermobility

But hypermobility also generates sizable social costs. Daniel Okimoto, a political scientist who studies U.S. and Japanese high technology, outlines some of these costs:

> Mobility is a hedge against the kind of rigor mortis that occurs when companies get locked into established technologies. . . . America's mobile labor force . . . allows companies to respond swiftly to new technological developments and market opportunities. But, high labor turnover exacts a high price in terms of information costs, transaction costs, potential legal entanglements, discontinuities in experience, loss of company investments in manpower training and the slowing down of momentum.[41]

Hypermobility, like the breakthrough economy of which it is a crucial constituent, creates the vexing condition of individual benefits pitted against "social" costs—costs that are passed on to other individuals, other firms, or the economy at large. These costs can be separated into four related categories: (1) the disruption of ongoing

R&D efforts, (2) a sacrifice of "institutional memory," (3) loss of investment and subsequent underinvestment in human resources, and (4) extreme career compression leading to high rates of worker burnout.

Disruption of R&D

Ongoing R&D is vital to corporate success, even survival, in high-technology fields. High rates of defection can seriously disrupt and weaken important R&D projects. The resignation of one or two employees is usually not devastating, but when three or more top-level employees in a specific area resign, the effect can be traumatic. And when entire project groups resign, the effect can be shattering. When people leave, they take their ideas and their know-how with them. Such capabilities may be very difficult and at times impossible to replace.

The severity of a departure depends on who and how many employees leave. For example, when Steve Jobs left Apple to form Next Computer, he took with him some of the top personnel from the then-completed Macintosh project. They included an Apple Fellow and top engineering designer, the marketing manager of the higher educational division, the manager of Macintosh software engineering, the senior controller for U.S. sales and marketing, and a very experienced engineering manager. This exodus not only affected Apple's technological capabilities in new product development, but it was also a serious blow to staff morale. Apple was able to recover because it was a big company with large reserves, but a smaller company might not have survived.[42]

Because R&D is so central to the operations of the small firms of the breakthrough economy, the near certainty of defections limits the ability to do long-term planning. The dilemma is simple: How do you plan for a project that can be thwarted by a few key defections? Defections require a firm to incur its "start-up costs" all over again. This is especially costly since R&D frequently takes place along a learning curve in which the most valuable learning and ideas come later. Large numbers of start-ups can result in a serious dilution of technical talent because, as in any cutting-edge field, there are only a limited number of top-notch personnel available. When each one is a member of an individual company, opportunities for synergistic interactions are limited.

Lost Institutional Memory

Hypermobility makes it difficult for a company to develop institutional memory, which is embodied in the cumulative expertise and knowledge possessed by its workers. When key people leave, the firm loses part of its memory, making it hard for it to grow and learn as an organization. In cutting-edge high-technology fields, such knowledge may be difficult, if not impossible, to replace. Since knowledge is frequently specific to a given firm or organization, it is impossible for new hires to function as perfect substitutes. And knowledge that was useful at one firm may not be relevant at another.

The mere threat of defections may cause firms to become oversecretive and adopt measures that weaken their ability to learn as organizations. A common strategy is to compartmentalize information so that workers only have limited knowledge of the product or process they are working on. While understandable as a defense measure, this only slows down the flow of information and impedes the interaction necessary both to innovate and to translate innovations into commercial products. Such policies may in fact destroy the organizational advantages that entrepreneurial start-ups now have over large U.S. corporations.

Human Resource Dilemmas

High rates of employee mobility make it necessary but also difficult for firms to make optimal investments in training and human resource development. High rates of mobility make it difficult for firms to get a full return from the investments they make in human resources. As a result, fearing that those investments may simply walk away, firms may underinvest in people.

Companies are caught in a difficult bind. One obvious strategy would be to hire an engineer, invest as little as possible in upgrading that engineer's skills, and then encourage that person to leave when those skills become obsolete. The cycle could then be repeated by hiring fresh recruits from the university. But companies are forced to invest more in human resource development, often just to get new workers up to speed and to compensate for skills that have been lost through defections. Moreover, new recruits demand "training" as a way to increase their own value. While larger companies like Intel and Hewlett-Packard can afford this, start-ups have a much more difficult time doing so. Seeq's Larry Jordan explains: "Employee training is a big part of our budget: it costs $25,000 to train a production worker in the first 6 months, and even if we're well under the industry average of 40% annual turnover, we still have a lot of training to do."[43]

The consequence of this is a baffling paradox: individual companies invest tremendous amounts in human resource development, but from a societal point of view these investments are not nearly enough to keep up with our competitors. This places the breakthrough economy at a serious competitive disadvantage with large Japanese companies, which can internalize their investments in human resources because they keep employees for their entire career.

Worker Burnout

Hypermobility burns people out, and quickly. As chapter 3 indicates, an intense and furious work effort is a key organizational element of the breakthrough economy. This takes a huge cumulative toll as workers move from one start-up to another—one pressure cooker to the next. One high-technology executive observes: "I put my three VPs in the hospital within the last two years, and three of the top four officers . . . lost their families through divorce in the past year. Suddenly you wake up and realize your kids are two or three years older, and one of them is in trouble. If it costs ten years of your life for every year you are president of the United States, each start-up in Silicon Valley costs you five years.[44]

The breakthrough economy leads to a staggering process of "career compression." The intense work environment of high-technology start-ups consumes think-workers very quickly. Careers can be as short as ten or fifteen years. The breakthrough economy trades the boredom and ennui of the corporate R&D laboratory for a high-pressure, high-burnout environment. In addition, some think-workers end up "falling through the cracks" of this hypermobile system. Participants in disasters may be branded "losers" or "high-risk propositions" and face difficulty raising venture capital in the future. Employees who leave large companies like GE or IBM may find it impossible to go back to their old jobs because such firms are often reluctant to rehire people who have been "polluted" by the start-up mentality.

The social costs of losing so many top-performing personnel are extremely high. Our system, which is premised upon an intense, high-motivation environment, threatens to consume its own "seeds"—skilled, inventive engineers. The U.S. can ill afford to sacrifice its top technical and managerial talent in the pursuit of short-term riches.

The Externalization of Innovation

Hypermobility is causing a sweeping transformation in the process of technological innovation itself. In the past, a steady stream of related innovations would be captured or internalized by single companies, making them ever more formidable.[45] Innovations today are far less likely to be captured by the companies that make them. For individuals and project teams, it is often easier and more lucrative to form a new company to develop an innovation than it is to develop it within an existing firm. Hypermobility makes it virtually impossible for existing companies to capitalize on the innovations they make, which gives rise to a phenomenon we refer to as the "externalization" of innovation. The nub of this process is summed up by William Poduska, founder of Prime Computer, Apollo Computer, and Stellar Computer: "The time to start a new business is when there is a real thunderbolt of a breakthrough. We hit it just right with Apollo; we hit it pretty good with Stellar."[46]

The externalization of innovation stems in part from the continuous nature of innovation process in high-technology industries. Small start-ups typically generate more technological opportunities than they can exploit. They lack the resources needed to develop new technologies that are not directly related to their own product line and frequently cannot commercialize all the innovations they come up with. These companies must spend all they can to upgrade and support existing products. But frequently, somewhere in the company there is a group exploring either the next wave of technical advance, which will supplant the existing technology, or another technology that is not in the mainstream of the company's product line. This places the company in a dilemma. It has created, but cannot support, the next round of technological evolution in its industry: it has sown the seeds of its own "creative destruction." Gordon Moore observes:

> Any company active in the forefront of semiconductor technology uncovers far more opportunities than it is in a position to pursue. When people become enthusiastic about a particular new opportunity but are not allowed to pursue it, they become potential entrepreneurs. As we have seen over the past few years, when these potential entrepreneurs are backed by plentiful sources of venture capital there is a burst of new enterprise.[47]

But the externalization of innovation would not be possible without the venture capital market, which provides a tremendous source of outside R&D funding. Phil Kaufman, president of Quickturn Systems, describes a situation commonly faced by the managers of our most innovative companies:

> One of my guys comes to me with a new idea and I can't finance it beyond a two-year time horizon. But he can go down the street and get three to five million in venture capital to launch his company. Venture capital represents a huge pool of R&D money. The problem is how can we figure out a way not to have every new idea develop into a new company. How do we push these ideas within existing companies?[48]

Some innovative start-ups have tried to respond to this situation by investing their own funds in spin-off companies. For example, both LSI Logic and Cypress Semiconductor have provided 100 percent equity financing for a series of new spin-off companies.[49] Although such investments may give these two companies a better "window" on the technologies being developed by their progeny, this is certainly not a solution to the broader problem.

The breakthrough economy is now faced with a situation in which existing companies find it virtually impossible to develop next-generation products or future technologies; that is, virtually every new innovation creates its own company. The once-virtuous circle linking innovation to new products and thereby increasing revenues and stimulating more R&D has given way to a vicious cycle in which a highly innovative firm stands a good chance of causing its own demise. Our high-technology future is increasingly entrusted to narrow "niche companies" that are unable to grow into something more. What's worse, these one-shot companies divide and dilute the base of talent and resources, weakening our overall ability to compete in high-technology fields. Even the largest, most stable high-technology companies are unable to escape the affliction, as their employees are lured away by huge financial gains afforded by start-ups and spin-offs. In this precarious environment, it is virtually impossible to build stable, integrated high-technology capabilities, to sustain long-term R&D efforts, or to follow through on innovations.

The externalization of innovation and the hypermobility of high-technology labor bring us face-to-face with a deeper, even more vexing dilemma, one that cuts to the heart of how we organize our leading high-technology industries. Many believe that the networks of small firms of Silicon Valley and Route 128 give the United States an unbeatable edge in the escalating global competition in high technology. Can this Silicon Valley and Route 128 model really save us? This is precisely the issue to which we now turn.

6

Silicon Valley and Route 128 Won't Save Us

■

There are a variety of things [high] technology industry depends upon that really can't be done well by small companies. Some of the major systems problems have such a broad scope that they require a fairly large group of mixed technical talents. . . . Many technological things have to be undertaken over a long time. After some of the initial work is done it can be relatively efficient to move to fruition through a start-up. But increasingly, it takes big investments and long times to do the basic technology.
—GORDON MOORE, Intel[1]

For many, if not for most, Americans, Silicon Valley and Route 128 stand out as symbols of economic and technological success. Their image of freewheeling, high-technology entrepreneurship and quick-shooting venture capital fits in nicely with our free enterprise ideology. These areas are typically held out as models for the rest of the American economy—in striking contrast to the failure of our large corporations and the economic devastation faced in older industrial regions.[2]

The reasons for the rapid growth of Silicon Valley and Route 128 are widely misunderstood. In recent years, two theories have received great attention for their ability to explain this model of high-technology organization and for their faith that this model can keep the U.S. ahead of its major competitors.

The first is the view that small firms are somehow better suited to new high-technology fields than are big ones. This view is most boldly formulated in George Gilder's "law of the microcosm," which suggests that small entrepreneurial firms have a natural advantage in the new "micro" technologies of the high-tech age.[3] He contrasts this with the old-scale economies that operated so well for the production of large things like automobiles and slabs of steel. In Gilder's words: "Rather than pushing decisions up through the hierarchy, microelectronics pulls them remorselessly down to the individual. This is the secret of the new American challenge in the global economy. . . . With the microprocessor and related chip technologies, the computing industry has replaced its previous economies of scale with new economies of microscale."[4]

The second theory suggests that networks or communities of small firms are a more effective form of economic and technological organization than are large integrated companies. The main proponents of this theory, Michael Piore and Charles Sabel, argue that "flexibly specialized" networks of small firms are characterized by close social relationships, shared trust, and intense cooperation in the development and production of new products.[5] They base their theory on case studies of the high-fashion clothing and footwear industrial districts of northern Italy, an area they refer to as the "Third Italy." According to this theory, Silicon Valley and Route 128 are high-technology versions of Europe's "cooperative" industrial districts where firms cooperate with one another in the development of new products, making it possible for them to remain small but still be globally competitive. According to a recent account by one of their students: "Silicon Valley firms describe their relations with suppliers in the language of personal rather than business relationships. They talk of building trust, making long-term commitments, 'holding hands with,' and even 'getting into bed with' suppliers."[6]

These theories offer an easy "things will take care of themselves" solution to the high-technology challenge—an imaginary high-technology world in which "microscale," "flexibility," "trust," and "cooperation" keep the U.S. economy ahead of the pack. But before we accept these answers, we must examine the evidence for these claims.

A Hobbesian World

The unfortunate reality of Silicon Valley and Route 128 is one of severe, at times devastating competition that drastically limits the ability of small entrepreneurial firms to cooperate with one another or to generate follow-through. Rather than a harmony of interests, the reality is one of each protecting his own—a trait clearly reflected in the recent rash of lawsuits charging companies with stealing employees or copying technology (see chapter 9). Cypress Semiconductor, a relatively recent start-up, currently faces at least 20 intellectual property lawsuits. Larger firms like DEC and Intel have developed in-house staffs of ten or more lawyers to deal with intellectual property litigation.

The Hobbesian side of American high technology is especially evident in the highly competitive relationships between companies. In the cutthroat environment of Silicon Valley and Route 128, passing problems on to others is considered smart business rather than a violation of trust. Each firm, its venture capitalists, and stockholding employees try desperately to increase their profits and their success at the expense of both their competitors and their "collaborators," for example, their suppliers. While a few large companies like DEC, Hewlett-Packard, and Apple have tried to develop closer, longer-term relations with their suppliers, most have not. In the pressure-cooker environment of Silicon Valley and Route 128, there is little burden sharing between companies; contracts are broken and suppliers let go when a better deal can be had elsewhere. A recent study of the state of the U.S. semiconductor industry is clear on this point: "The U.S. semiconductor producers and their equipment and material suppliers are disaggregated and have little tradition of cooperation and mutual support."[7]

Most companies try to drive prices down by pitting one supplier against another. Not surprisingly, suppliers respond to these conditions by competing mainly on price, delivering cheaper products that are of lower quality and are less reliable. And if a shortage arises, suppliers have every incentive to drive up prices mercilessly. According to *Electronic Business:* "Electronics manufacturers, like most others, have traditionally used a common approach when choosing goods and services suppliers: give us your cheapest prices and we'll give you our business."[8]

Stockpiling and hoarding are commonplace in the high-stakes environment of Silicon Valley and Route 128, where companies try to outguess the market and where suppliers seem to rise and decline overnight. Apple Computer, a company hailed as a pioneer in close supplier relations, lost tens of millions of dollars in 1989 when it stockpiled dynamic random access memory (DRAM) semiconductor chips in anticipation of shortages and escalating prices, as recounted in a recent report: "After hoarding millions of dollars worth of expensive DRAMs, when the market was tight, chip supplies loosened and the market fell. Alas, the personal computer maker was stuck with a stockpile of overpriced memories—a costly purchasing blunder. . . . Apple's dilemma is not unique. A single misguided purchasing decision can save or lose millions of dollars for a company."[9] This inefficient strategy of stockpiling also injects incorrect information on demand into the market.

Most high-technology companies of Silicon Valley and Route 128 make little effort to develop tight permanent relationships with even their most valued suppliers. The "arm's length" relationship is the rule. Companies may register a complaint when deliveries are late, but that is the extent of it. They generally do not foster any communication beyond the actual purchase agreement. Suppliers may be asked their "opinion," but they are rarely seriously consulted on design options. In fact, most current attempts to reform supplier relations are more hype than reality. DEC's much-heralded "Key Supplier" program extends to just 20 of the company's 2,000 plus suppliers. According to Ron Payne, vice-president for corporate purchasing at DEC, the company is "moving away from a straight competitive bid environment and toward a longer term relationship" for this select group of key suppliers.[10] But for the rest, competitive bids and arm's-length relationships are the order of the day. Hewlett-Packard's new supplier program depends mainly on punishing suppliers who deliver defective products or make late deliveries. For example, when a defect is identified, all Hewlett-Packard divisions are placed on "quality alert" and prohibited from buying from the supplier until the defect is corrected and a "correction notice" is issued.[11] These so-called reforms stand in sharp contrast to the long-term, mutually supportive supplier relations found in Japan where large hub firms work closely with their suppliers to help them solve quality or delivery problems.

What's more, Silicon Valley and Route 128 firms often go outside their region to find low-cost sources of supply. A recent study based on detailed interviews with 40 high-technology companies in Silicon Valley indicates that more than half the companies have relationships with equipment suppliers located outside the region and that these relationships are mainly commercial or "arm's length" in nature. Furthermore, roughly two thirds of the principal components or inputs used in the development of new products come from suppliers outside the region. Local suppliers mainly provide "nontechnology" products, such as cabinets, casings, power supplies, raw materials, basic chemicals, and office supplies and highly standardized high-tech components, such as computer disks. The study concludes that: "the region's integration within a broader global milieu is increasing as local linkages . . . decline in importance."[12]

Silicon Valley and Route 128 firms contract out a large and growing share of manufacturing to specialized "contract manufacturers" chosen mainly on the basis of price (as opposed to proximity or quality). Contract manufacturers frequently locate their plants in the Sunbelt and Third World where labor costs are low. A common pattern is to have a main sales office and a small plant in Silicon Valley or Route 128 to serve special customer demands and larger, high-volume plants in low-wage Sunbelt states, Asia, and Mexico. Flextronics, a leading contract manufacturer for Silicon Valley semiconductor firms, has plants in the Sunbelt, New England, Southern California, and Hong Kong, as well as Silicon Valley. SCI Systems, another important contract manufacturer, is headquartered in Alabama to take advantage of low wages. As we will see, a large and growing number of high-tech companies have manufacturing done by foreign companies, such as the Taiwan Semiconductor Manufacturing Company, who pay extremely low Third World wages (see chapter 7).[13] Contract manufacturers offer a cheap outlet for production but seldom if ever perform the role of collaborative partners in the development or improvement of products.

Much attention has been focused on the cooperative relationship between Sun Microsystems, a cutting-edge workstation manufacturer, and Cypress Semiconductor, a leading custom chip company that produces tailor-made chips for Sun.[14] But the close relationship between Sun and Cypress is an exception engineered as an "experiment" by the venture capital fund Kleiner Perkins, which backs both companies. According to John Doerr of Kleiner Perkins, the venture fund has sought to re-create elements of Japan's industrial network structure among the entrepreneurial companies in which it has invested. Kleiner Perkins wants to play the dual role of central financial institution and anchor institution for a galaxy of linked corporations, a role that in some ways resembles that of the bank in the Japanese *keiretsu* system.[15]

Moreover, Sun has experienced problems in its attempts to fashion a model "solar system" of vendors and suppliers. According to a recent report, many current members of Sun's solar system are constantly worried by their dependency on their "Sun," which could at any time move to cut them out.[16] Furthermore, Sun was unable to find a suitable American supplier for the original reduced instruction set chip (RISC) that forms the heart of its pioneering workstation. After a series of failed negotiations with American chip makers, Sun turned to the Japanese electronic giant Fujitsu for this product.[17] Cypress came into the picture later. In a Fujitsu advertisement in *Business Week,* Sun president Scott McNealy is quoted as saying: "This is our longest steady corporate relationship. Fujitsu has been a key partner of ours. . . . Fujitsu was one of the first companies to take a chance on Sun. They treated us like a big company when we were barely out of the start-up phase. They have supplied us everything from DRAMs to disk drives."[18]

The Sun-Fujitsu relationship highlights another drawback to the Silicon Valley/Route 128 model of the high-technology organization: so many American high-technology companies buy their products from Japanese multinationals rather than from local "partners." The director of purchasing for a U.S. high-technology company provides some perspective: "The Japanese concept of just-in-time is to ring your plant with suppliers in the shadow of your building. That concept is fine but it is in basic conflict with the fact that in the U.S. many of our suppliers are in the Pacific Rim."[19]

CEOs and purchasing directors of U.S. start-ups provide countless stories of how they are unable to get their domestic suppliers to provide high-quality components and are forced to turn to Japanese competitors instead. Robert Shillman, CEO of Cognex, a Route 128 company, indicates that his company was willing to pay a 20 percent premium to a U.S. supplier, but still ended up buying from a Japanese company because no U.S. firm could match the quality.[20] Jerry Crowley, chairman of Gazelle Microelectronics, a Silicon Valley manufacturer of gallium arsenide chips, says that Kyocera's San Diego plant was one of the first to do business with his small 11-person company, which many U.S. companies considered too small to do business with.[21] Don McCranie, CEO of Seeq, says his company has no choice but to turn to a Japanese supplier for semiconductor packages (the finish work on semiconductors) because "there are no U.S. manufacturers left capable of making semiconductor packages. . . . Now if you buy a [semiconductor] package, you buy a Kyocera package. Every missile system made in the United States may have a U.S.-made chip in it, but it will be a Kyocera package."[22]

To serve this growing market, a growing number of Japanese firms are opening state-of-the-art semiconductor design centers in Silicon Valley and the Route 128 area. These firms work closely with their American customers to come up with tailor-made designs. The custom designs are then dispatched via electronic mail to state-of-the-art Japanese manufacturing facilities that produce the actual chips, which are then shipped back to the U.S. by air. These Japanese firms are now going the next step by building new semiconductor manufacturing facilities in the U.S.

Leading U.S. semiconductor companies are becoming overwhelmingly dependent for their production equipment upon Japanese companies, such as Canon, TEC, and Nikon. LSI Logic, for example, gets more than 60 percent of its production equipment from Japan because domestic manufacturers are unable to deliver reliable, high-quality products.[23] Certainly, if relationships were harmonious and highly interactive, Silicon Valley's semiconductor companies would not have abandoned their fellow U.S. suppliers of production equipment. Indeed, a history of strained, adversarial relations between chip companies and semiconductor equipment manufacturers underscores the decline of the U.S. semiconductor production equipment industry.[24]

On yet another dimension, American producers of advanced semiconductor design automation equipment and high-tech instruments are finding that large Japanese corporations comprise a large and rapidly growing share of the market for their cutting-edge products. Cognex not only depends on Japanese suppliers, but sells between 20 and 40 percent of its machine vision systems in Japan.[25] MRS Corporation, a leading manufacturer of x-ray lithography equipment used in advanced "active-matrix" laptop computer screens, currently sells 80 to 90 percent of its systems in Japan and the rest in Europe; the company has been unable to sell any in the U.S.[26] Executives of leading Silicon Valley design automation firms like Silicon Compiler Systems, Cirrus Logic, Synopsys, and SDA (now Cadence) and high-tech equipment instrument producer KLA Instruments told us that Japanese corporations are buying more and more of their products.[27] Even advanced semiconductor manufacturers like Brooktree and Altera are doing more business in Japan; Altera estimates that by 1991 more than 20 percent of its sales of programmable logic chips will come from Japan.[28] According to *Electronic Business,* foreign sales by the top 200 U.S. high technology electronics companies rose by 42 percent in 1988, while domestic sales increased by a mere 2.4 percent.[29]

The point of all this is basic: While Silicon Valley and Route 128 firms may innovate locally, the markets for their high-technology products are shaped by strong market forces that are increasingly global in scope and may in fact ultimately contribute to the demise of U.S. high technology itself. Harvey Jones—president of Synopsys, former president of Daisy Systems, and former venture capitalist—sums it up succinctly: "You cannot build a high-technology economy by flipping out start-ups and leave the rest of it to the Japanese."[30]

Start-up Mania

The Hobbesian side of the U.S. model of high-technology organization is further reflected in the proliferation of "me-too" start-ups, or copycat companies, that occupy nearly all areas of the high-technology industry. Me-too start-ups are linked to "technology fads," the rapid rise of hot new technologies. In the high-stakes world of American high technology, the emergence of a new technology produces a rush of clones as everyone tries to cash in on the latest technology fad. Donald Valentine explains: "The sopping up of resources by multiple startups . . . is detracting from the competitiveness of U.S. industry. It used to be that the only competition we faced was from larger, well-established companies that didn't recognize a market niche or an opportunity. It took most of us to finance one company into business, two at the most. Now . . . each [venture capital] group feels it has to have one of every kind of an investment.[31]

The phenomenon of me-too start-ups is related to the tremendous expansion of the venture capital pool, which increases both the opportunity and the pressure to produce new start-ups. John Wilson uses the term "feeding frenzy" to convey the reckless abandon in financing clone start-ups.[32] An excellent example of such overinvestment is the personal computer disk drive industry:

> From 1977 to 1984, venture capital firms invested almost $400 million in 43 different manufacturers of Winchester disk drives . . . including 21 startup or early stage investments. . . . During the middle part of 1983, the capital markets assigned a value in excess of $5 billion to 12 publicly traded, venture capital backed hard disk drive manufacturers. . . . However, by 1984, the value assigned to those same 12 manufacturers had declined from a high of $5.4 billion to only $1.4 billion. . . . When viewed in isolation each [funding] decision seems to make sense. When taken together, however, they are a prescription for disaster.[33]

The result was a shake-out of companies with large financial losses and massive layoffs. A second domestic shake-out that occurred in 1989 left only four U.S. drive makers—Conner Peripherals, Maxtor Corporation, Quantum Corporation, and Seagate Technology—to meet the growing challenge of large Japanese companies who are rapidly increasing their share of the hard disk market. Quantum has in fact established a partnership with Matsushita to manufacture low-end disk drives. Conditions have gotten so bad for U.S. disk drive producers that George Scalise of Maxtor has called for the establishment of a federally supported consortium of U.S. disk drive companies to overcome the damaging effects of domestic competition and beat back the Japanese challenge.[34]

In biotechnology, similarly, the proliferation of start-ups has been great, with venture capital funds backing over two hundred new companies. Numerous start-ups scrambled to be the first to market in nearly every product category.[35] This is a new development in the pharmaceutical industry, which has a strong tradition of patenting and "first mover" advantages. While the biotechnology industry has thus far avoided a shake-out (mainly by opting for mergers among small companies), me-too start-ups have spread the narrow base of biotechnology talent across a large number of companies and caused considerable redundancy in R&D. This has also forced a host of start-ups into joint ventures with domestic and foreign competitors.

Me-too start-ups, in dividing market share and talent among companies, weaken many in ways that can threaten the development of entire industries. The proliferation of copycat companies in narrow business areas makes it difficult to establish the continuity it takes to follow through and often leads to serious misallocations of resources, business failures, and destabilizing shake-outs. Thus, clone companies may appear rational from the perspective of each entrepreneurial group and venture investor, but they often end up hurting the high-technology industry as a whole.

The Hobbesian realities of Silicon Valley and Route 128 fly in the face of academic theorists who would like to explain their technological dynamism and economic performance in terms of a theory of economic cooperation based on the high-fashion clothing and footwear industries of northern Italy. Perhaps the most insightful perspective on this issue was provided in an interview we conducted with Luigi Mercurio, an Italian high technologist and former Olivetti executive who now lives and works in Silicon Valley as CEO of David Systems.[36] Mercurio sees little if any similarity between Silicon Valley and Italy's much-heralded industrial districts. For him, Silicon Valley epitomizes a free-wheeling, entrepreneurial economy where technological innovation is motivated by the potential to profit and accumulate great wealth. The "Third Italy" exemplifies an "old-world economy" where generations of family ties exert powerful influence over the local economy. In Silicon Valley, the rule of profit dominates; the firm itself has become a commodity to be bought and sold to the highest bidder. In the "Third Italy," decades-old social ties and community relationships place strict limits on economic behavior and the "family firm" remains a source of livelihood and support for many generations. One does not need a theory of cooperation and trust to explain the innovativeness and economic dynamism of Silicon Valley or Route 128, when a simple understanding of the super-profits that come from breakthrough innovation will do. Although a cooperative community of high-technology firms and their suppliers is certainly a desirable ideal, the reality is a competitive war of all against all in which the outcome is tremendous profit for some and exhaustion for many others.

Industrial Fragmentation

All of this has contributed to a growing problem of industrial fragmentation, which comprises two elements, one horizontal and one vertical. These can be illustrated by a tree metaphor. A firm's basic product can be considered the trunk. Successful R&D not only strengthens the trunk but also leads to the development of new branches from the trunk, that is, new products in related areas. The branches allow the company to diversify and become stronger and more stable. Horizontal fragmentation transforms these branches into self-standing companies. The parent company is left with its trunk and very little possibility of growth through branching. Vertical fragmentation splits the tree into separate sections running from the roots to the leaves. Instead of working together as a single organism, the tree is split into several independent entities, that is, roots, branches, and leaves. Each of these segments then must operate in its own best interests to ensure its own profitability.

The semiconductor industry is the most obvious case of horizontal fragmentation. The semiconductor industry is split into five segments: merchant producers, captive producers, integrated circuit producers, design specialists, and subcontract manufacturers.[37] According to one recent report: "America's semiconductor makers are mostly specialist, independent companies; Japan's are high volume subsidiaries of giants. In the past five years, America's small firms have lost their dominance of the world memory chip market to big Japanese rivals. They now fear they will lose the rest of the business too."[38]

Horizontal fragmentation leaves most semiconductor firms entirely dependent upon a single core product. A narrow product base makes it impossible to cross-subsidize products, leaving many firms vulnerable to major technological changes, price swings, and industry downturns. A slump in personal computer or workstation sales, for example, could wipe out a whole host of specialized semiconductor producers and even cause significant financial problems for large producers. And when an international price war breaks out, such highly specialized companies are in trouble almost immediately. High degrees of industrial fragmentation have weakened the U.S. semiconductor industry, leaving it increasingly unable to respond to foreign competition.

The computer industry is divided into at least ten separate segments, with only a handful of companies like IBM, DEC, and Hewlett-Packard important players in more than one. The mainframe segment remains dominated by IBM, and to a lesser extent by the remaining members of the BUNCH [i.e., Burroughs, Univac (now Unisys), NCR, Control Data, and Honeywell (now out of the business); two Japanese players, Fujitsu and Hitachi, have also joined this sector], as well as "plug compatible" manufacturers, like Amdahl.[39] DEC, Data General, IBM, and HP are the major producers of minicomputers.[40] Fault-tolerant computers are made by Tandem, Stratus, and Tolerant Systems. Sun Microsystems and Hewlett-Packard are the largest producers of engineering workstations.[41] Apple, IBM, and Compaq dominate the personal computer segment.[42] Laptops are made by Compaq, Grid Systems, and Zenith, and smaller notebook computers are being made by new start-ups, including Agis, Go Corporation, Information Appliance, and Poqet Computer, which is 38 percent owned by Fujitsu.[43]

Within this industry, the supercomputer segment has undergone an extreme horizontal division. It is populated today by more than twenty-five companies, which compete in a variety of areas.[44] The company most usually associated with supercomputing is Cray Research.[45] Recently, Cray has been challenged by Supercomputer Systems, a company founded by a former Cray computer scientist, Steven Chen, and forty-five defecting Cray employees.[46] However, the supercomputer segment has divided into a series of minisegments. A new minisupercomputer segment has emerged and is dominated by two new start-ups, Convex and Alliant, but recent entrants include Elxsi, Floating Point Systems, Multiflow, Saxpy, SCS, Sky, Gould, and Cydrome.[47] Parallel processing computers that use more than one processor are being made by new entrepreneurial companies like Sequent Computer Systems, Encore Computer, Thinking Machines, Flexible Computer, Floating Point Systems, NCube, and BBN Advanced Computers.[48] Manufacturers of desktop supers include Ardent Computer and Stellar, which merged in 1989 under the aegis of the Japanese firm, Kubota.[49] The tragedy in supercomputers is that these companies do not communicate and share information either across or within segments.

Obviously even IBM's formidable presence has not prevented massive waves of new entrants and increasing fragmentation in the computer industry. According to computer industry expert, Kenneth Flamm: "IBM sometimes lagged in the introduction of new technology into its product line: time sharing systems, the use of integrated circuits, large-scale supercomputers, small-scale minicomputers and microcomputers, and software making use of artificial intelligence are areas where IBM trailed more aggressive competition."[50] For a time IBM was able to absorb these new developments and cope with, indeed capitalize upon, increasing fragmentation. But as we have seen, even "Big Blue" has come under increasing pressure both from start-up companies and from Japanese competitors in recent years.[51]

American high technology suffers from an extreme form of vertical as well as horizontal fragmentation. Vertical fragmentation means that various functions of the firm, ranging from R&D to manufacturing, are parceled out to independent firms, where each aspect of the production chain is the province of a separate group of specialized companies.

The production of customized chips is a case study in vertical fragmentation. Custom chips (application-specific integrated circuits, or ASICs) are designed by specialized design firms, produced by independently owned foundries, and assembled by still another group of companies. The measurement, test devices, and other equipment used in the production process are made by yet other firms. In fact, only a handful of the recent semiconductor start-ups, such as LSI Logic and VSLI Technology, are integrated producers with complete design and manufacturing capabilities. Others such as Brooktree, Cirrus Logic, Maxim Integrated Products, S-MOS Systems, and Xilinix are "fabless" companies that have no manufacturing capability or fabrication plant and are entirely dependent upon outside manufacturers.[52] Gordon Bell explains that the future of the U.S. semiconductor industry may be a "completely segmented industry in which the user, designer, . . . design center, and foundry are all separate."[53]

For some, like George Gilder and T.J. Rodgers of Cypress Semiconductors, this new configuration is heralded as evidence of the flexibility and renewed competitiveness of the U.S. semiconductor industry.[54] But their optimistic spirits are dampened by the rise of large Japanese corporations who have become the main players in the low end of the customized chip business and who are squeezing the profit margins of LSI Logic and VSLI Technology, the most important U.S. producers of custom chips. An increasingly uncompetitive brand of overspecialization, not flexible specialization, is the distinctive feature of the U.S. semiconductor industry.

Vertical fragmentation also exists between industries. Consider the fault line between the semiconductor and computer industries. Only a handful of our largest computer companies, including IBM, DEC, and Hewlett-Packard, make the chips that go into their products; the others depend on outside merchant suppliers. In fact, IBM is the only major U.S. computer manufacturer that remains a major force in the development of new chips. According to Dataquest, IBM produced $1.8 billion of the $2.9 billion in semiconductors it consumed in 1986, purchasing the rest from outside suppliers.[55]

When Intel recently moved to vertically integrate by producing personal computers and computer workstations based on its own microprocessors, it sent shock waves through the American high-technology community, essentially calling into question the long-held division of labor between semiconductor producers and the computer systems makers they supply. Intel's move provoked an outcry from a number of the company's leading customers in the computer industry who see it encroaching on their business and has even caused some to actively search out new suppliers. Compaq Computer, one of Intel's most important customers, has retaliated by providing venture funding for NexGen Microsystems, a company that produces clones of the Intel microprocessors used in its computers.[56]

This example illustrates a basic point. The highly fragmented structure of U.S. microelectronics has created patterns and rules of behavior that make it extremely difficult for companies to integrate. Companies like Intel that try to integrate or otherwise break away from the existing structure run the risk of alienating their customer and/or supplier base and may seriously jeopardize their own financial condition. While U.S. semiconductor and computer firms are engaged in this frantic jockeying for position, large, integrated Japanese corporations continue to make greater and greater inroads in virtually every microelectronic market from mass-produced semiconductors to custom chips and from laptops and personal computers to high-end supercomputers.

Taken together, horizontal and vertical fragmentation produce a pattern of industrial development that is the opposite of the traditional pattern of vertical integration, whereby large numbers of entrepreneurial firms eventually give way to large, integrated enterprises.[57] This fragmentation of American high technology did not just happen; it was the result of conscious historical choices. The choices and successes of the founding fathers of American high technology created a set of institutions that supported entrepreneurial high technology and made fragmentation appear "natural." Fairchild and Intel founder Robert Noyce explained the historic choice made in the American microelectronics industry and, by extension, in American high technology: "We are going to less and less vertical integration. . . . All electronics firms do not feel that they must make their own semiconductor devices; nor do they feel they must grow single crystals, make their own masks, build their own furnaces or test equipment."[58]

Fragmentation can damage entire high-technology complexes, which tend to specialize in a narrow band of high-technology products. The dramatic growth of Route 128 during the late 1970s and early 1980s was driven by the increasing demand for minicomputers, dedicated word processors, and office automation products manufactured by DEC, Data General, Wang Laboratories, Prime Computer, and other Route 128 companies. However, the shift to "distributed" personal computers that began in the mid-1980s undermined the market base of the manufacturers of these minicomputer and dedicated office systems and helped create a regional recession in 1989.[59]

The fragmentation and splintering of our high-technology capabilities makes it ever more difficult to build stable companies and industries that can compete over the long haul. Even our strongest, most innovative companies are finding it difficult to grow and prosper in such a highly fragmented environment. The extreme segmentation of the high-technology production process drastically inhibits follow-through and hinders American industry's ability to meet the challenge of emerging global competition.

Innovation Dilemmas

The combination of small size and industrial fragmentation makes it difficult for American high-technology firms to combine one or more technologies into new hybrid innovations or to generate systems technologies. "Mechatronics," the combination of mechanical and computer technologies, is a good example of a hybrid technology.[60] Mechatronic products include consumer goods like watches, cameras, and home appliances and industrial goods like industrial robots and machine tools. Systems technologies come from the combination of a variety of technologies in a workable system.[61] Television, telephones, and electrical transmission are good examples of systems technologies. For television to be successfully commercialized, a wide range of products such as television tubes, cameras, receivers, and transmission equipment had to be combined into a workable system.

Hybrid innovations and systems technologies can be of even greater economic importance than radical new product breakthroughs. The application of mechatronics by Japanese firms to wrist-

113

watches, for example, revolutionized the watch market, opening up a huge new market in inexpensive and reliable quartz watches. According to recent reports, large Japanese companies dominate many important new hybrid fields, including mechatronics and "optoelectronics," the combination of computer and video technologies.[62] Our weakness in high-temperature superconductivity, which involves the combination of electronics, computers, ceramics, and manufacturing technologies, provides another telling example of U.S. weakness in hybrid innovation.[63]

High-definition television (HDTV) illustrates our weakness in systems technology. HDTV promises to make current television systems obsolete and open huge new markets for microelectronic products and applications. The global market for advanced television systems alone is expected to reach $30 billion by the year 2000, and go as high as $500 billion by the year 2020. Many believe that HDTV is a critical "enabling technology" with important ramifications for a host of technology fields and industries. HDTV will revolutionize the home entertainment industry, creating new markets for video compact discs, laser printers and turntables, video libraries, and even "computerized" television.[64] These products provide an enormous demand for an entire spectrum of electronics components. It has been estimated that the demand for memory chips for HDTVs could be five times larger than the total demand from the computer industry.[65] HDTV will fuel a host of related innovations in fields like display technology, imaging systems, medical diagnostics, and even radar systems. And sales of HDTV products will provide the capital needed to undertake huge investments in new digital communications infrastructures such as fiber optic lines, which can handle increased electronic data loads. Hugh Carter Donahue observes:

> Cable and VCR transformed television in the 1970s and 1980s, providing unheard of flexibility in programming. High-definition television promises to have an even greater impact by the 1990s. HDTV sets will show high resolution pictures on large, extra-wide screens and will produce the crystal clear sound of a compact disc. They may also be smart enough to store and retrieve electronic still pictures, allow two way video communication, and receive pro-

gramming from broadcasters, cable, satellite, and perhaps even fiber optic phone line.[66]

But the United States is already behind in the race to HDTV, as large Japanese companies like Sony, Matsushita (Panasonic), Toshiba, and Hitachi consolidate their lead in television and video electronics. Steven Jobs sums up the current condition of the U.S. HDTV effort: "All this stuff about how the U.S. is going to participate is a joke. . . . We've lost it already."[67]

The reasons for our failure in hybrid innovations and systems technologies are easy to understand. The small high-tech companies of Silicon Valley and Route 128 lack the scale, resources, and long-term outlook that are needed to develop these types of products. When companies make just one version of a product or produce just one part of a product, they have neither the breadth of in-house expertise necessary to create important hybrid innovations nor the large numbers of R&D personnel necessary to undertake a large systems innovation. In the words of Regis McKenna: "Small companies are great product innovators, but they have limited resources. They can initiate innovation, but few can sustain it."[68]

The Missing "Consumer Connection"

The problems of our HDTV effort illuminate a critical issue facing American high technology: the huge chasm that separates the innovative high-tech firms of Silicon Valley and Route 128 from our traditional consumer products industries.

This is a two-way street. On the one hand, few leading U.S. high-technology companies make consumer electronics goods.[69] Small entrepreneurial companies that have tried to enter mass production fields have usually failed. For example, when Intel and Texas Instruments tried to get into radio and digital watch production, they were quickly annihilated by large Japanese companies. On the other hand, large electronics companies, such as GE and Westinghouse, are certainly not important producers of cutting-edge commercial high technology. Most of the high-technology products produced by these companies are for military applications. As discussed in chapter 2, large U.S. electronic companies would rather produce for the lucrative defense sector than make commercial products.

There is little connection between small high-technology firms and large consumer electronic companies, so they do not reinforce one another's activities. Consider the following facts. Just 6.4 percent of all U.S. semiconductor sales are to consumer electronics companies.[70] The vast majority of semiconductors are consumed by other high-technology sectors, for example, computer and telecommunications firms and the military. The result is that the American semiconductor industry is left with a narrow and volatile market base. American microelectronics as a whole misses out on the potentially huge profits and reinvestment capital that can come from using high technology to make better consumer products. American high technology is faced with a missing "consumer connection."

This missing connection is even more disastrous for the consumer electronics industry. This chasm makes it difficult to apply and use new technological developments to upgrade and improve older products. As a consequence, the U.S. consumer products industry has fallen far behind those of Japan and Western Europe. The U.S. Semiconductor Industry Association recognized this problem in 1989: "Our international competitors are far ahead of us in developing advanced electronics applications via the consumer market segment. We, as an industry, and as a country, must pull together quickly . . . before our competitors have an insurmountable lead. . . . America's future industrial viability and economic leadership could well be at stake."[71] The situation is so serious that a presidential commission on the semiconductor industry issued a report in October 1989 calling for the establishment of a new multibillion-dollar Technology Corporation of America to resurrect the consumer electronics industry.[72]

Large Japanese companies are able to use microelectronic innovations to develop cutting-edge consumer electronic goods such as Watchman televisions, CD players, miniature tape players and recorders, and a wide range of other products. But Japanese corporations are also using advanced technology to revolutionize the "white goods" industry by applying high technology to everyday needs such as cooking, keeping food cold, and washing clothes. Take the example of a simple standardized commodity, the home washing machine. Japanese washing machines use chips to replace mechanical parts. As a result they are quieter, more reliable, and less expensive to produce than American washing machines, most of which are still made with mechanical gears (see chapter 8).

In biotechnology, the missing consumer connection is also evident—but in a different way. Start-ups work on narrow niche technologies but lack the marketing networks for the products they develop. Large companies are unable to secure the best researchers and thus must purchase marketing rights to fill out their existing product lines. For this reason the new techniques of biotechnology are not adequately integrated into the knowledge base of the firm. In addition, a variety of promising uses for biotechnology ranging from food processing, mining, and waste cleanup either go unaddressed or move along at a slow pace.

Enclaves of Restructuring

Perhaps the most striking shortcoming of Silicon Valley and Route 128 is that their model of technological and industrial organization has done little to transform basic manufacturing industries like consumer electronics or automobiles. In fact, the major organizational innovations associated with this model are virtually unheard-of in large Fordist industrial corporations. It is as though high-technology start-ups and large industrial companies are operating on entirely different planets.[73]

This is apparent in the strained relationships between microelectronics companies and the automobile industry.[74] High-technology entrepreneurs and the straitlaced managers of General Motors and Ford think on entirely different wavelengths. Rapid technical change and the ability to make fortunes overnight create a contemptuous attitude among high-technology executives for what they consider a backward "Rust Belt mentality." Automobile executives, in turn, have a distaste for the idiosyncrasies of "hotshot" entrepreneurs. The constant strife between H. Ross Perot, founder of Electronic Data Systems, and Roger Smith of General Motors was just one example of this. According to Robert Palmer, an expert on relationships between the automobile and semiconductor industries, "automotive buyers attempt to treat the electronics as a rustbelt industry. . . . There's more stated partnership than real. . . . Semiconductor companies tend to send signals that they have a lack of patience with metal benders. They have a technical arrogance."[75] The lack of trust and communication between high-tech firms and Big Three automobile companies means that American cars contain a relatively low level of microelectronics, and what microelectronics they do possess is less than state of the art. Our own study of component parts suppliers to the Japanese "transplant" automakers in the American Midwest reinforces this conclusion: Japanese companies have experienced great difficulty getting microelectronics parts from U.S. producers and continue to import them from Japan.[76]

It is naïve to think that the model of high-technology organization found in Silicon Valley and Route 128 can save us from the challenge of heightened global competition. While this model gives rise to new, highly innovative companies at breakneck speed, it also generates a high degree of internal competition and a serious problem of industrial fragmentation. It can catalyze the world's most advanced breakthrough technology, but it is unable to generate the small-product, process, hybrid, and systems innovations that are needed to turn breakthroughs into a wide variety of commercial products. In the end, Silicon Valley and Route 128 remain two limited enclaves of restructuring that have been unable to transform the main body of the U.S. economy either through the diffusion of their organizational practices or by setting in motion the "gales of creative destruction" that can reinvigorate and renew traditional industries. Even though the breakthrough economy can find rich veins of technological opportunity, it is unable to mine those veins fully.

7
Neglecting High-Technology Manufacture

■

We don't care very much about manufacturing.
—GORDON CAMPBELL, president, Chips and Technology[1]

American companies don't like to build things, they like to make deals.
—GORDON BELL, leading computer designer[2]

Go visit a high-technology manufacturing facility in Silicon Valley or Route 128 or further afield in Malaysia or Puerto Rico. What will be most striking is the way engineers, R&D scientists, and other think-workers so easily stand out from production workers. The engineers and managers are in total control, giving orders that production workers obey. Production workers do not attempt to engage engineers but simply go about doing their work. One is witness to the strict division between those who think about, plan, and manage work and those who actually carry out the work.

This reflects a central principle of the breakthrough economy: that only scientifically oriented breakthroughs count, and that only a small group of innovators provide the source of new ideas and hence new economic value.[3] The president of a leading semiconductor start-up puts it as follows: "Manufacturing is no longer an art. The real value added is in architecture, design, and testing."[4]

As a consequence, our leading high-technology corporations neglect, even ignore, manufacturing. Manufacturing is little more than a necessary nuisance. One commentator offers this explanation: "Especially in the small electronics industry companies, manufacturing is

119

the last thing people think about. It is viewed as a cost center and generally receives management attention only when problems occur."[5]

Consider the following facts. Production workers now account for less than 20 percent of the roughly 200,000 employees of Silicon Valley high-technology firms. In 1988 there were less than 40,000 high-tech production workers in Silicon Valley compared with more than 100,000 high-tech professionals, managers, R&D workers, and salespeople.[6] And according to estimates released in January 1990, U.S. high-technology electronic corporations expect to increase their investment in manufacturing plant and equipment by just 2.3 percent over the course of the year compared with increases of 10 to 15 percent for Japan, Hong Kong, South Korea, and Taiwan.[7]

The entrepreneurial start-ups of the breakthrough economy thus fall into the same trap as the large Fordist firms of the old follow-through economy. Factory workers are seen as source of sweaty labor—a view that ignores the ways in which they can contribute to the innovation process. Unfortunately, this view is shared by many Americans who take comfort in the belief that the United States has ascended to the position of a postindustrial nation in which manufacturing is unimportant. This philosophy may have worked when their was no competition, but it cannot work anymore.

Workers as Second-Class Citizens

There are two tiers of workers in most high-technology enterprises: highly paid think-workers who work in gleaming office parks and an almost invisible stratum of low-paid factory workers who perform repetitive, unskilled tasks under the guidance and authority of the think-workers. The state of California's standard job description for electronics assembly workers indicates that such workers must have an ability to perform repetitive tasks, "an ability to follow oral instructions," and "the ability to sit in one place for long periods of time."[8] This worker profile is hardly aimed at securing involved and committed employees.

High-technology production workers are predominantly women and members of minority groups who do not share in the benefits afforded to the largely male think-workers. A recent report based on data collected by the U.S. Equal Employment Opportunity Commission shows the high degree of work force segregation and dualism in Silicon Valley's high-tech work force. In 1988, more than half of unskilled operatives, one third of laborers, and three quarters of clerical workers were women, but women held only 20 percent of management jobs. (White men constituted two thirds of management and 56 percent of the professional work force). Minorities made up the great majority of high-tech production workers. More than 70 percent of all production jobs were held by members of minority groups: 43 percent were Asian; 21 percent Hispanic; and 6 percent African-American. Just 16 percent of high-tech managers were minorities. The report qualifies this figure concluding that: "ethnic Japanese, Chinese and Indians appear to prevail in high level positions, while Filipinos and Vietnamese make up the greatest number of Asian production workers."[9]

While high-technology companies have created highly interactive and participative environments for their engineers and managers, the organization of manufacturing work appears as if it is deliberately designed to stultify thinking. This is clear from the following description of a typical high-technology electronics factory: "Unlike auto factories, steel mills, and other manufacturing facilities, semiconductor and computer manufacturing plants generally do not have large numbers of employees performing functions on an integrated line or manufacturing process. The work is done in small units with six to thirty workers. Jobs are compartmentalized and closely supervised, with a relatively high ratio of supervisors to workers."[10]

High-technology companies use a variety of mechanisms to control production workers and limit their discretion in the work process. There is little knowledge transfer among shop-floor workers or between production workers and managers. Although production workers are in constant contact with the production process, they are not allowed to act on their own judgment but must wait for direction from supervising engineers.[11] Workers in contract manufacturing plants in Silicon Valley and the Sunbelt face even lower wages and worse working conditions. Conditions for Third World assembly workers are even worse. Their work is highly repetitive, speedup is common, and they have virtually no discretion in the performance of their tasks. The overriding concern of management is maximizing physical output, not the quality or improvements that come from knowledgeable and committed workers.

Low Pay

Pay is a market indicator of the value placed upon various types of work and workers. And pay for high-technology production workers is quite low. In 1988 entry-level pay for unexperienced electronics assemblers ranged from $4.75 to $8.00 an hour, while those with more than three years experience received between $7.00 and $10.50 an hour.[12] At the low end of the scale are temporary workers; at the high end are employees of large companies like Hewlett-Packard.

This is a far cry from the $40,000 plus salaries afforded entry-level engineers, and it is shocking when compared to the multimillion-dollar remuneration packages of top-level management. *Electronic Business* reports that the 1988 salaries for the hundred highest paid electronic executives averaged $821,000, without stock options or bonuses—fifty to one hundred times what production workers make.[13] John Sculley, president of Apple Computer, topped the list at $2,479,000. Ten other executives made over $1 million and the entire top hundred made more than $500,000 dollars each. Taking into account stock options and other nonsalary compensation, a number of executives received compensation of over $10 million. The raises top executives receive to switch companies can be just as astounding. In June 1988, for example, Joseph Graziano resigned as chief financial officer of Sun Microsystems and returned to his previous employer, Apple Computer. To lure him back, Apple paid Graziano a $1.5 million signing bonus and agreed to pay him as much as $600,000 in annual salary and bonuses. This was quite an increase from his 1988 remuneration at Sun—a mere $237,000 in salary and bonuses. Supposedly, Graziano did not even ask Sun to match his offer "but simply resigned and left the company the same day, even though he was putting together a huge financing."[14] In stark contrast are those working on the production line or in other semiskilled positions—generally women or members of minority groups with the greatest needs—who do not participate in the financial benefits afforded to largely male think-workers.[15]

Poor Working Conditions

Aspects of the work environment provide another indicator of management's view of the importance of manufacturing. Health and safety conditions in high-technology factories can be quite poor, for instance.[16]

A report by Joseph LaDou, a specialist in occupational medicine, provides alarming statistical evidence of the serious hazards faced by high-technology production workers.[17] According to LaDou's statistics, which are based on data from the California Department of Industrial Relations, the rate of occupational illness for workers in semiconductor factories is three times higher than that for all manufacturing workers—1.3 illnesses per hundred workers compared with 0.4 per hundred workers among all workers. Semiconductor workers are three times more likely to fall victim to serious job-related ailments that result in lost work time. Eighteen percent of semiconductor workers miss work as a result of occupational illnesses and injuries as compared with 6 percent for all manufacturing workers. And nearly half (46.9 percent) of the job-related illnesses that afflict semiconductor workers are from "systemic poisoning"—prolonged exposure to toxic materials. That is twice the rate for all manufacturing workers.

Job Insecurity

High-technology production workers have very insecure employment conditions.[18] Annual turnover rates for line workers, board stuffers, and assemblers often range from 30 to 50 percent.[19] High-tech production workers find little reason to stay with a firm because layoffs and forced departures are common. Such workers are usually the first to be laid off in a downturn; foremen and manufacturing-related personnel are only slightly more secure. For example, in response to the 1974 recession, Intel idled almost a third of its work force. One Intel employee described the layoff process this way:

> I was a line supervisor. My boss came to me and said that I would have to lose all the people in my group. . . . Fifteen minutes later, I was back in my office thinking that was really unpleasant but thank goodness it didn't get me. My boss called me in and asked me if I'd taken care of the layoffs. I said yes, and he said 'I'm really sorry to tell you this, but you yourself are now laid off.' So fifteen minutes later I was out in the parking lot talking to the . . . guys that I had just laid off. Fifteen minutes later, here came the plant guards with my boss. . . . All of us really cared about the company and wanted to be working there. We were real people with real lives. But now we didn't have a job.[20]

Insecure employment conditions are not just the workers' problem; they are a problem for the breakthrough economy as a whole. Insecure and fleeting employment makes it hard for firms to build up a skilled and knowledgeable work force. Workers in turn develop little loyalty to their employers. Only a handful of companies, such as Hewlett-Packard, offer employment security and try to avoid layoffs by shortening the work week or instituting across-the-board pay cuts during downturns. While other companies have flirted with employment security, most of these efforts have been short-lived. After introducing a "no layoff policy" in 1981, Intel instituted massive layoffs in the wake of the 1986 downturn in the semiconductor industry.

Even companies with long-held commitments to permanent employment are beginning to lay off workers under increased competitive pressure from domestic as well as Japanese firms. In April 1990, the *Wall Street Journal* reported that IBM, DEC, and Hewlett-Packard all reversed their "no-layoff" policies and began to implement work-force reductions. IBM plans to slash employment by more than 10,000, while DEC will cut between 5,000 to 8,000 employees mainly through early retirement and severance programs. Hewlett-Packard laid off 440 workers from its newly acquired Apollo unit, stating flatly that: "It's impossible today to have a cut-and-dried no-layoff policy."[21] For most high-technology companies, even large, successful ones, manufacturing workers are expendable and becoming more so.

Full-time production workers in large manufacturing facilities are only the tip of the iceberg. Many companies use temporary workers to fill manufacturing positions. For example, in 1986 there were roughly fourteen thousand temporary workers in the Silicon Valley semiconductor industry alone.[22] As discussed later in this chapter, there are also a large number of subcontractors that do manufacturing work on a short-term contract basis. Temporary workers face low pay and very insecure conditions of employment. They function as a safety valve for high-tech manufacturing operations, since they can be brought in and let go at the company's whim. As expendable "transients," they are offered few benefits, with little or no investment in their training.

Nonunion Jobs

Only a small percentage of high-technology production workers have the security of union employment. In 1987 the unionization rate for the electronics industry as a whole was just over 20 percent, well below the 59 percent rate for the automobile industry or the 48 percent in the steel industry.[23] According to Steve Early, a labor journalist, and Rand Wilson, a former high-tech union organizer, only ninety companies in the American Electronics Association had union contracts, leaving approximately 2.5 million electronics workers without union representation.[24] In the small firms of the biotechnology industry, there has been virtually no unionization at all.

The "human resource" practices and production organization of many high-technology firms are designed specifically to keep unions out. Some of the larger companies, like IBM, have policies providing internal job ladders, regular wage increases, and employee associa-

125

tions or unions to avoid unionization. Others simply move plants to nonunionized locations in the Sunbelt or Third World. The use of large numbers of temporaries and part-time workers fosters divisions in the work force that make it difficult to form unions.[25] The following account shows how high-technology plants organize production and work schedules in ways that inhibit worker solidarity:

> Plants tend to employ fewer than 500 workers, and products are manufactured in a cluster of facilities that may be miles apart. Furthermore, many companies require some employees to work a standard eight-hour day five days a week while others work four ten-hour days or three twelve-hour days. With staggered shifts, only a portion of the work force is in the plant at any one time. Such arrangements and the use of temporary, part-time, and week-end-only workers make it difficult to forge the solidarity necessary for a successful unionization campaign.[26]

High-technology workers who do try to organize unions or even air grievances may be fired. In one case, a worker was fired from National Semiconductor simply for circulating a petition in support of another worker who had been arbitrarily terminated. In another, a Signetics worker was fired for supporting a union. The National Labor Relations Board ruled in favor of the workers in both cases.[27] This resembles the extreme antiunion hysteria common among industrial manufacturers at the turn of the last century.

The mere threat of unionization has caused some high-technology companies to move production overseas. Atari is a case in point. According to industrial relations specialist Everrett Kassalow, when Atari's initial approach of small wage increases and a new "family approach" to labor-management relations backfired and the plant was on the verge of unionization, Atari simply moved its production facilities to Taiwan and Hong Kong, eliminating some twenty-five hundred jobs in California.[28] Atari workers then filed a National Labor Relations Board suit, and after years of legal wrangling, the company agreed to an out-of-court settlement.[29] But the plant and the jobs were gone.

Moving Production to the Third World

Driven by the opportunities for savings in labor costs and visions of happy, docile workers, a large and growing number of our leading high-technology companies have moved a substantial percentage of their assembly and manufacturing activities to Asia, Mexico, and the Caribbean. The movement of high-technology manufacturing to Third World countries is a natural extension of the philosophy that production and production workers are only a necessary nuisance.

Today an enormous amount of high-tech assembly and manufacturing takes place in the Third World. According to a recent study, there are ninety-one foreign semiconductor plants, sixty-three of which are in Southeast Asia.[30] More than 90 percent of all semiconductor assembly work is done overseas. As a result, U.S. semiconductor companies have more foreign production workers than domestic ones. Motorola is currently building a $600-million, 326,000-square-foot semiconductor manufacturing plant in Hong Kong's Silicon Harbor Center. This state-of-the-art plant will give the company complete manufacturing and design capabilities in Hong Kong. According to Tam Chung Din, director of Motorola's Hong Kong operations, the Hong Kong plant is more advanced than any of Motorola's U.S. facilities.[31] A 1988 Dataquest survey indicates that roughly half (46 percent) of semiconductor firms plan to move more factory production abroad in the future.[32]

The computer industry is also distinguished by a penchant for Third World manufacturing. In recent years DEC, Wang, Apollo, and numerous other firms have opened up Korean branch plants (see table 7.1). A large percentage of the "all-American" IBM personal computer is manufactured overseas. The same thing is evident in the disk drives and peripherals industries. Seagate Technology, the world's largest producer of Winchester disk drives, has moved nearly all of its production to Southeast Asia.[33] These are then assembled in Asia or returned to the largely automated U.S. assembly facilities.

A recent survey of leading electronics corporations by Ernst & Young provides even more striking evidence of the dimensions of offshore manufacturing.[34] Semiconductor, communications, and computer companies make the most use of foreign manufacturing plants. More than one-third of all semiconductor companies and one-quarter of computer and communications manufacturers have offshore factories. And the great majority of larger high-tech firms operate offshore plants. According to the survey, nearly three-quarters (72 percent) of companies with revenues in excess of $300 million and almost two-thirds (61 percent) of firms with revenues between $100 and $300 million have manufacturing plants located offshore. On the whole, more than 20 percent of all high-technology electronics companies operate manufacturing plants outside the United States.

This reality remains hidden from many Americans, because so many of the final products bear American names. According to *Business Week:* "The severity of the situation is masked by the fact that U.S. brand names continue to dominate critical markets, such as computers and scientific instruments—although many of their components are turned out by foreign workers in foreign factories."[35] But this does not change the fact that most of the jobs and manufacturing wealth is created outside the United States.

TABLE 7.1

U.S. Branch Plants in South Korea

Company	Investment	Year	Product
U.S. Owned			
Texas Instruments	$13,800,000	1988	Integrated Circuits
Litton	10,000,000	1986	Electronics
Western Digital	3,500,000	1988	Boards
DEC	4,000,000	1988	Computers
Wang Computer	1,500,000	1986	Computers
Apollo	200,000	1986	Workstations
Microsoft	100,000	1986	Software
U.S.-Korea Joint Ventures			
Monsanto/Lucky	6,300,000	1986	Silcon Wafers
ELCO/Hanukilbo	5,500,000	1987	Connectors
EDS/Lucky Goldstar	2,100,000	1987	Software
Convex/Seoul Electron	400,000	1987	Computers
EPRO/not available	100,000	1987	Semiconductor Equipment

Source: Dataquest, "The Electronics Migration to Korea: Who and Why," *Dataquest Research Bulletin* (June 1988); reprinted with permission.

The movement of high-technology manufacturing to Third World countries has a long history. Semiconductor manufacturers began transferring assembly operations overseas during the 1960s. During that decade alone, American companies established fifty overseas branch plants, perfecting a locational strategy referred to as "island hopping."[36] Wafers, fabricated domestically, would be shipped overseas to low-wage sites where they were cut and processed and then returned to the U.S. or sent to other markets. In 1962 Fairchild opened its first foreign manufacturing facility in Hong Kong; it followed with a plant in Korea in 1965, another in Singapore in 1967, and one in Indonesia in the early 1970s. Its offspring, Intel, opened a component assembly and testing plant in Malaysia in 1972, and another in the Philippines in 1974. As Asia became more expensive, Intel shifted production to the Caribbean, opening an assembly plant in Barbados in 1977 and another in Puerto Rico in 1981. Then, while the company laid off U.S. workers in 1984, it opened another foreign plant in Singapore.[37] Between 1963 and 1985 the share of American workers engaged in direct production of semiconductors declined from 66 to 40 percent.[38]

The breakthrough economy has come full circle. The tremendous overseas migration of manufacturing has re-created the same international division of labor—the same separation of innovation and production—that was pioneered by large industrial corporations of the older follow-through economy.[39] In the words of Donald Valentine: "Silicon Valley and Route 128 are worlds of intellectual property, not capital equipment and production. Most of the employees of U.S. high technology live in Southeast Asia."[40] In 1985, for example, U.S. semiconductor corporations employed 150,000 foreign factory workers and just 115,000 domestic production workers.[41]

And some high-technology companies have taken this international division of labor one step further, moving their midlevel product development or specialized R&D to countries like Taiwan or India, which have a large supply of low-wage engineers. In the last two years IBM has established two new R&D subsidiaries in Taiwan, Neotech Development Corp. and International Integrated Systems, to develop new printer components and integrated circuits.[42]

High-technology companies locate factories in the Third World

129

for the same reasons large industrial companies do: low wages, union avoidance, an absence of occupational health and safety legislation, and a lack of environmental regulations. To encourage U.S. firms to move abroad, some Third World countries have even set up liaison offices in Silicon Valley and other parts of the United States, which provide information and facilitate the securing of necessary approvals and overcoming other obstacles.

But nothing lasts forever. The once-docile environments of many Third World nations have turned into hotbeds of worker unrest. Third World workers are asking for decent wages and better working conditions; and in some countries they are forming new industrial unions to advance their demands. Wages are clearly on the rise. Between 1984 and 1988 pay increased by 110 percent in Korea, 84 percent in Taiwan, 50 percent in Hong Kong, and 32 percent in Singapore.[43] Ironically, these wage increases have been spurred mainly by the massive expansion in overseas manufacturing itself, which is the underlying reason for tighter labor markets and even labor shortages. Strikes and work stoppages are also on the upswing. During 1988 and 1989 serious labor disputes erupted at more than fifty foreign-owned plants in Korea, causing production setbacks and delivery problems. In response to this, Tandy Computer and Fairchild Semiconductor recently closed their embattled Korean plants. Instead of elevating manufacturing work and the conditions of manufacturing workers to humane conditions, U.S. companies continue to scour the globe for new sources of cheap labor and ever more controllable environments. And in doing so, they only make their already very serious manufacturing problems worse.

The Subcontracting Explosion

Perhaps the clearest indication of this neglect for high-technology manufacturing is the explosion in subcontracting—a strategy that allows companies to ignore manufacturing completely. A growing number of new start-ups do not engage in manufacturing at all. These are referred to as "fabless start-ups" because they operate without a semiconductor manufacturing facility, which is referred to as a fabrication plant. According to *Electronic Business:*

> The fabless semiconductor company is not a new phenomenon. For years, Silicon Valley has witnessed a sprouting crop of companies that have shunned manufacturing. Fledgling firms like Altera Corp., Chips and Technologies, Inc., Vitelic Corp., and Xilinix Inc. are but a few examples of U.S. chip start-ups that have concentrated on the design side of the business, leaving their manufacturing for others to do. At first these fabless start-ups were mocked. "Real semiconductor companies have fabs" sneered some in the industry. But if imitation is the truest form of flattery, the start-ups now have reason to feel smug. Many of the major U.S. chip makers—AMD, Intel Corp., and National Semiconductor Corp., chief among them—confess that they are now farming out more and more of their manufacturing. In fact, Intel will contract out between 20 and 30 percent of its total production.[44]

Foreign subcontracting is a rapidly growing facet of the breakthrough economy. This trend is especially pronounced in the semiconductor industry, where the Asian share of subcontracted manufacturing is as high as 60 percent, according to some estimates. Foreign subcontractors, typically referred to as "foundries," do silicon wafer processing for customers like Intel, AMD, National, Motorola, and a host of smaller semiconductor producers. In this new international division of labor Texas Instruments has work done by Japan's NMB Semiconductor and Korea's Hyundai; Motorola has semiconductors made by Toshiba; National Semiconductor uses NMB and Hyundai; Intel uses Mitsubishi Electric and Samsung; and AMD uses a variety of Japanese firms. According to the most recent available estimates, between $500 million and $1.2 billion worth of semiconductor manufacturing is contracted out to foreign firms each year.

Computer companies are also becoming increasingly dependent upon foreign subcontractors. Between 1985 and 1988 Apple Computer, IBM, Sun Microsystems, Unisys, and Bendix all signed subcontracting agreements with Korean manufacturers such as Samsung and Goldstar.[45]

The full dimensions of the foreign subcontracting explosion are outlined in the Ernst & Young survey of high-tech electronics companies.[46] According to this survey, 42 percent of all communications firms, 40 percent of computer manufacturers, and 37 percent of semiconductor companies rely on foreign subcontract manufacturers. Both large and small high-tech firms contract out manufacturing to foreign companies. Roughly 42 percent of larger high-tech firms (those with revenues in excess of $100 million) and one-quarter of small high-tech firms (those with revenues less than $25 million) have manufacturing done by foreign subcontractors. All in all, nearly one-third of all high-tech electronics firms depend on foreign contractors for manufacturing.

The tremendous upsurge in foreign subcontracting makes it even harder to determine how much "American" high technology is produced at home and how much is produced overseas—a troublesome phenomenon aptly dubbed "the new shell game" by *Electronic Business.* [47]

Companies choose to subcontract out manufacturing for a variety of reasons. For some companies, especially young ones, subcontracting is a strategy that allows them to pour money into high-end R&D and avoid capital-intensive investments in manufacturing plant and equipment that drain huge amounts of financial and human resources. While venture capital provides enough cash for companies to get going, it is not nearly enough at current prices to construct state-of-the-art integrated manufacturing facilities. New semiconductor factories cost between $200 and $350 million to build. VSLI Technology's new $75 million "minifab" has captured the attention of semiconductor industry observers who hope this will spur a wave of reinvestment in American semiconductor manufacturing. Nevertheless this is still a substantial investment, larger than many new start-ups are willing or able to make.[48] When the tradeoff is between R&D for the development of innovative new products and manufacturing that can be farmed out, manufacturing inevitably loses.

For others companies, subcontracting can provide a buffer against swings in the business cycle. As chapter 6 indicates, high-technology companies see little value in building long-term cooperative relationships with their subcontractors. Subcontractors are looked upon as easily replaceable. Steve Pletcher, vice president of a growing independent foundry, observes that "foundries and assembly houses have often been considered a safety valve: If I have to ramp up quickly, I will ramp up my subcontractors. When I have to ramp down quickly, I will ramp down my subcontractors."[49]

For still others, subcontracting is an explicit strategy to reduce direct labor costs and/or inhibit workers' efforts to organize unions. In 1984, for example, Intel closed eight of its own plants and began to increase its use of outside subcontractors. This resulted in mass layoffs of some six thousand workers and managers—nearly 25 percent of its work force.[50] David House, an Intel vice president, outlines this strategy: "We've been increasing our subcontracting over the last two or three years. It used to be two or three percentage points and now it's over 10%. We might not stop at 30% but our objective is not to go to 100%."[51] Subcontracting takes the separation of the production and innovation to a new extreme, with manufacturing taking place in organizations that are unrelated to those that innovate.

The Hollowing of High-Tech Industry

Taken together, the rise of offshore manufacturing and growing dependence on subcontracting is leading to the "hollowing" or "deindustrialization" of high-technology industry similar to what happened in old-time manufacturing industries like steel and automobiles.[52] Simply put, the United States is abandoning its high-technology manufacturing base.[53] Akio Morita, Sony's cofounder and chief executive, observes that "American companies have either shifted output to low-wage countries or come to buy parts and assembled products from countries like Japan that can make quality products at low prices. The result is a hollowing of American industry. The U.S. is abandoning its status as an industrial power."[54]

The adverse effects of hollowing are reflected in the dire state our high-technology manufacturing effort. A recent study by Hambrecht and Quist indicates that our semiconductor yields are now three times lower than those of Japanese companies.[55] According to Dataquest, between ten and twenty U.S. semiconductor fabrication plants have closed each year since 1986. Since openings of new plants have not kept up with closures, the result is a net reduction in our capacity to manufacture semiconductors. We have fallen so far off the cutting edge of semiconductor facility construction that an increasing share of new American semiconductor fabrication plants, including IBM's new advanced chip facility in East Fishkill, are being built by Japanese companies.[56]

Will U.S. firms be able to compete if they are unable to manufacture their innovative products? The answer to this question does not bode well for the breakthrough economy. It is impossible for an economy that neglects manufacturing to remain competitive over the long haul. James Koford of LSI Logic explains:

> In high-technology fields, the U.S. is rapidly becoming a non-manufacturing nation. We sell our innovations and get a one-shot infusion of capital, not a continuous product stream. Manufacturing is the ability to make a lot of things, it is the engine which drives progress. If we lose this base, we don't have the economic engine to fuel innovation. We need a complete infrastructure to continue to get good innovations. We need good manufacturing capabilities, an emphasis on quality, good engineering schools. We need to tweak science and technology to three orders of magnitude through our manufacturing capabilities.[57]

The problem runs deeper still. Hollow companies—companies that become dependent upon outside manufacturing—suffer a variety of adverse consequences. Most of all, such companies are unable to create the much-needed interplay between manufacturing, design, and R&D. It is difficult, if not impossible, to integrate these critical activities successfully when manufacturing is located offshore. Constantinos Markides and Norman Berg suggest that "[a] business cannot design in a vacuum. It cannot exploit new technologies, if it has no chance to apply them. . . . The fact is, design and manufacturing are linked. A company that subcontracts its manufacturing to foreigners will soon lose the expertise in design and the ability to innovate because it won't get the feedback it needs."[58]

Hollowing creates other, less obvious, problems as well. U.S. corporations that employ foreign subcontractors run the risk of creating their own competitors. David House of Intel suggests as much: "You really have to ask yourself: Long term am I creating more monsters. It concerns me whether our [foreign] foundries will remain happy in their role as just manufacturers or if one day we'll see them as across-the-board competitors."[59]

Subcontractors learn much about the products they make, since their customers must transfer technology in the form of blueprints, product specifications, machinery, and even engineers to assist with manufacturing setup and quality control. This makes it easy for subcontractors to "reverse engineer" and clone the products they manufacture. Japanese companies that manufacture components for our leading computer and semiconductor firms learn a tremendous amount about that state-of-the-art technology.

The Japanese are a classic instance of subcontractors becoming competitors and displacing their former patrons in the marketplace. Now the Koreans and Taiwanese intend to follow the Japanese example.[60] For precisely this reason, Hyundai, for example, actively seeks semiconductor subcontracting even when its own production lines are booked up. By 1988 Korea had become a major site for the processing, production, and assembly of semiconductors—ranking third behind Japan and the United States, but already ahead of West Germany. Increasingly, companies like Samsung, Goldstar, Daewoo, and Hyundai are important producers of a wide array of semiconductor products.[61]

U.S. high-technology firms have fallen so far behind in manufacturing and production technology that they are now being forced to establish manufacturing partnerships with Japanese corporations to gain access to state-of-the-art Japanese production technology and management techniques. In a highly publicized deal, Advanced Micro Devices (AMD) recently sold one of its Texas semiconductor factories to Sony for $55 million. As part of the deal, Sony will upgrade this plant and use it as a base to teach AMD executives and engineers about Japanese production technology. AMD will be able to transfer Sony's production technology to its other U.S. plants for five years. In another recent deal, Intel has established a joint venture with

Japan's NMB Semiconductor to market the Japanese company's memory chips and to acquire NMB's production technology for use in Intel's manufacturing plants. In the words of a top Intel executive: "One big thing we get out of this is to have them help us with automated factories."[62]

Allowing foreign firms to do our industry's manufacturing is clearly helping to prepare them for future moves into more advanced, higher-value areas. Even if a firm or a nation starts out with imitation, the experience soon breeds new ideas for products and processes and before long the capacity to innovate itself. When this happens, our high-technology leaders will have no one to blame but themselves.

The Crux of the Problem; or, the Separation of Innovation and Production Revisited

The breakthrough economy is confronted by a basic paradox of the high-technology age. The central importance of highly advanced technology to microelectronics, biotechnology, and other new industries makes it seem as though high-end R&D is the center of economic value and profit and that manufacturing is less important than ever. But the reality is that manufacturing is of even greater importance in these new industries than it was before.

The reason for this is rooted in the changing nature of technological innovation itself. In one of the most insightful examinations of the changing nature of contemporary industrial society, Tessa Morris-Suzuki advances the concept of "perpetual innovation" to explain the rapid and continuous nature of technological change that follows the shift from older mass production industries to new information-intensive technologies and industries.[63] This can be seen in fields such as personal computers, where state-of-the-art products become antiquated in two or three years; or semiconductors, where cutting-edge technology is outmoded even quicker, in a year or two. Basically these knowledge-intensive products and innovations are amenable to continuous upgrading and refinement. In this environment, the ability to improve products and processes constantly, to revamp the production process itself, and to deploy new products and technologies rapidly is critical. And since these process and product innovations are themselves quite knowledge-intensive, the nature of production itself is changing from a process based on the extraction of marginal units of physical labor to a process based on continuous knowledge extraction and innovation both in the lab and on the shop floor.

In this new age, only a state-of-the-art manufacturing capability can create the synergy with R&D that is needed to turn innovations into products as quickly as possible and to introduce improved, next-generation products rapidly into the market. Success requires a great deal of high-intensity communication and face-to-face interaction between these two critical activities; the time pressure to get products to market first makes such interaction absolutely critical. But the continued, increasingly global separation of innovation and production in our leading high-technology firms and industries makes this impossible to achieve. The results are problematic: long delays turning innovations into products, lower-quality products, and great difficulties keeping up with the competition as foreign companies begin to manufacture "clones" and then improved next-generation products.

To remedy this problem, U.S. high-technology firms need to build tight linkages between R&D labs and factories, between the sites of innovation and production. While they can and should start by relocating factories near R&D centers, simply bringing factories back home is not enough. There must be a change in those organizational practices and beliefs that imply that manufacturing is somehow less important than high-end innovation. Basically, the organizational separation of innovation from production must be overcome: the entire spectrum ranging from R&D to manufacturing must be turned into a seamless web. Tight connections between manufacturing and R&D are needed to enhance both our capacity to innovate and our capacity to follow through on the innovations we make.

In an era of perpetual innovation, manufacturing itself is a more important source of innovation than ever before. Factory workers possess a huge storehouse of practical knowledge and technical know-how that cannot be replicated in the R&D lab. This knowledge is absolutely necessary to identify and devise effective solutions for a host of problems; and it is an invaluable source of product and process innovation as well. Neglect of manufacturing cuts off two important types of innovation: small improvements in products that can be made during the production stage and important advances in manufacturing processes themselves. Both are critical in the current reality in which products and processes change quickly.

A constant theme of this book has been that the U.S. has created an entire stratum of employees whose potential contributions to creating value has been narrowly circumscribed and even discouraged. Even the new high-technology firms that were so effective in providing managers and engineers with the new participative environments needed to be creative and make a full contribution failed to do this for shop-floor workers. A critical failing of the breakthrough economy lies in the organization of manufacturing, that is, on the shop floor.

The new age of perpetual innovation requires a shift in both the organization of manufacturing and the role and responsibility of manufacturing workers. To compete in cutting-edge high-technology fields, it will be necessary to tap the intelligence of all workers, especially shop-floor workers. They must be seen not simply as "hands" but as complete persons who can think as well as execute. This means the extension of the participative environment afforded R&D scientists and engineers down to include shop-floor workers so that all workers become think-workers. Without this, the breakthrough economy and the broader U.S. economy of which it is a part will remain in their current problematic condition.

III

∎

BEYOND THE
BREAKTHROUGH
ILLUSION

8

Competing with Japan: The World's New Follow-through Economy

■

I can name a dozen areas where we used to laugh at the Japanese and now they are caught up. . . . Our lead is probably not recoverable except in damaged ways. We have lost the battle and I see no mechanisms for stopping it. We probably have been beat. We must develop a coexistence strategy.

—JAMES SOLOMON, founder and president, SDA[1]

In recent years an initially surprising, then bewildering, and now numbing litany of reports, studies, and forecasts of Japanese successes in field after field have been issued. It has become apparent that Japan has altered the rules of the competition that Europe and the U.S. have found so stable and reassuring since World War II. Naturally, this has spurred an equally mind-boggling avalanche of theories for why this is so. For many, the answer lies in the unfair play of Japanese government and industry—what the political scientist Chalmers Johnson calls the "developmental state"—in organizing, promoting, and protecting key industries.[2] For others, the key is to be found in the "bigness" of its high-technology corporations, which is the main source of its advantage over the U.S.[3] And for still others, Japan's advantage stems from a legacy of "hard work"—a regimen of long working hours, a fast work pace, and the discipline imposed by its leading corporations.[4] Much of what they say is true. Certainly, government and public policy have played an important role in Japan's technological and industrial development. Its leading corporations are indeed big, and there is no doubt that Japanese managers, engineers, R&D scientists, and factory workers work very hard.

141

But the key to understanding Japan lies in its revolutionary new forms of organization that generate a powerful ability to follow through—to turn innovations into a continuous stream of high-quality mass produced products. And this in turn stems from new organizational structures that enable Japanese corporations to harness the knowledge and intelligence of workers across the entire production chain in the R&D laboratory and on the factory floor. According to recent studies, Japanese corporations convert new innovations, especially those that come from outside sources, into products more quickly than do their American counterparts, compressing the time it takes to move innovations from R&D to manufacturing.[5] They can supply the needed improvement innovations for superior next-generation products. They are able to integrate new high technologies to make revolutionary advances in mass-market consumer products. And they excel at combining innovations to make hybrid and systems technologies. A Japanese chemist spells out an important element of Japan's follow-through approach: "Real industrial innovation is more than scientific breakthroughs, it spans the search for new materials, process technology, successful manufacturing schemes, and successful marketing."[6] To put it simply: Japan has become the world's new follow-through economy.

Size, Scale, and Synergy

Large companies are the cornerstones of Japan's follow-through economy.[7] A comparative study of innovation in Japan and the U.S. found that large Japanese companies accounted for all but two of thirty-four major innovations in Japan, whereas large U.S. companies accounted for just half of all major innovations in the U.S.[8] And Japan's large corporations have the size, scale, and synergistic capabilities to turn their new innovations into globally competitive products.

Japan's leading microelectronics companies now number among the largest and most successful in the world (see table 8.1). Japanese corporations include six of the top ten global semiconductor producers: NEC, Toshiba, Hitachi, Fujitsu, Matsushita, and Mitsubishi.

TABLE 8.1

Top Ten Companies in Semiconductors, Computers, and Telecommunications

Semiconductors		Computers		Telecommunications	
1986 Production	($ Millions)	1986 Sales	($ Billions)	1986 Sales	($ Billions)
1. NEC	$2,560	1. IBM	$47.6	1. AT&T	$10.2
2. Toshiba	2,270	2. Unisys	9.4	2. Alcatel N.V.	8.0
3. Hitachi	2,160	3. DEC	8.4	3. Siemens	5.6
4. Motorola	1,960	4. Fujitsu	6.0	4. N. Telecom	4.4
5. Texas Inst.	1,850	5. NEC	5.6	5. NEC	4.0
6. Phillips	1,325	6. Hitachi	4.6	6. Ericsson	3.1
7. Fujitsu	1,145	7. NCR	4.2	7. Motorola	2.8
8. Matsushita	1,145	8. Hewlett-Packard	4.2	8. IBM	2.7
9. Mitsubishi	990	9. Olivetti	3.7	9. Phillips	2.0
10. National	970	10. CDC	3.3	10. GEC	1.9

SOURCE: Alden Hayashi, "NEC Takes the Triple Crown in Electronics," *Electronic Business* (September 15, 1987): 140.

Three of these companies, NEC, Fujitsu, and Hitachi, also rank among the world's ten largest computer companies. NEC, though much smaller than IBM or AT&T, recently took the *Electronic Business* triple crown in high-technology electronics, ranking first in the sales of semiconductors, fifth in computers, and fifth in telecommunications. No American company, not even IBM, is this diversified across the information technology fields.[9]

But size alone is only part of the reason for Japan's tremendous economic success. Japan's leading high-tech corporations have pioneered a new form of organization based on synergistic growth, whereby the firm is constantly expanding into related fields. So, for example, the same Japanese companies that make semiconductors also make computers, telecommunications equipment, electronic instruments, office automation equipment, and industrial robots as well as mass-market consumer electronics goods.[10] Together, NEC, Hitachi, Fujitsu, Toshiba, Matsushita, and Mitsubishi Electric account for between half and two-thirds of all semiconductors, integrated circuits, computer, and computerized machine tool sales in Japan.[11]

The constant quest for new technologies and business opportunities is astonishing. For example, on a research trip to explore the R&D efforts of some of Japan's leading electronics and biotechnology companies one of the letters requesting an interview with a biotechnology company went to a major electronics company by mistake. Still, the interview went off without a hitch. The electronics company executives delivered an elaborate presentation on an entire portfolio of intriguing bioelectronics projects they were working on.[12]

This combination of size and scale generates important advantages. It provides a financial cushion that can be used to cut prices, introduce loss leaders to capture new markets, and stay the course during economic downturns. An example of this is the "leaked" Hitachi Memorandum that called upon its semiconductor salesmen to cut prices of memory chips by 10 percent until they made the sale.[13] It also creates tremendous internal markets for high-technology products, which can be increased or decreased depending upon outside market conditions, thereby providing an important buffer from the market. According to a recent report by the U.S. Semiconductor Industry Association:

The semiconductor divisions [of Japanese corporations] can often count on other divisions of the corporation to consume up to 35 percent of their output. . . . [I]ntegration gives . . . financial resources far beyond those of their American counterparts, provides them with a ready domestic market for their components, offers them leverage to encourage reciprocal buying from òther domestic companies, and allows them to design chips specifically for the upstream products their companies make.[14]

Synergy provides advantages that go beyond size and scale. Japanese corporations are constantly using their newest technologies to revolutionize mass-market consumer goods. For example, large electronics corporations adapt advances in semiconductor technology to improve televisions, stereo equipment, refrigerators, and washing machines while simultaneously using them to develop new office products and industrial machinery. A most unusual development is a new high-tech toilet that tests the user's blood pressure, pulse, and urine and electronically stores the results for up to 120 days.[15] Gene Gregory, an expert in Japanese high technology, explains:

In the process of diffusion, Japanese integrated circuit manufacturers have a special and often decisive advantage in the diversification of downstream equipment manufacture—including a wide variety of computers, communications equipment, robots, medical equipment, office equipment, home appliances and audio-visual equipment—which make possible rapid and extensive integrated circuit applications.[16]

A similar situation is occurring in biotechnology, where the breadth and integration of Japan's effort provides numerous outlets for new applications. Susan Clymer of the American firm Bio-Response comments in this regard: "I see a much higher rate of diffusion of biotechnology throughout Japanese industry than I see here in the U.S. In the U.S., I don't see such a creative view."[17]

Japan's large corporations can also achieve synergistic benefits from a single innovation. A striking example is video display technology: innovations that were originally used to develop small Watchman televisions are also used in new screens for portable laptop computers;

or high-resolution computer graphics displays are employed in new high-definition television (HDTV) systems.[18]

This method of organizing the firm also provides a huge advantage in hybrid and systems technologies.[19] Japanese companies are now pacesetters in mechatronics fields such as flexible manufacturing systems, industrial robots, computer vision systems, and new semiconductor fabrication techniques.[20] The same is true in optoelectronics, the fusion of optics and microelectronics technologies. Fusing optics with other technologies, Japanese companies are developing new types of integrated circuits, medical lasers, compact disks and disk memories, switches, text scanners, and even computer networks.[21] In another hybrid field, photonics, the combination of lasers and manufacturing technologies, a recent National Research Council report concluded that the U.S. has been reduced to an "observer."[22] And as we have already seen, Japan is already ahead in both superconductivity and HDTV, two of the most important new technologies of the 1990s.

But most of all, the combination of size, scale, and synergy provides both the resources and the scope that are needed to undertake further rounds of technological development. It allows corporations to cross-subsidize R&D and product development and to sustain low profits while gaining market share. And in doing so, it enhances their ability to make new innovations, to turn those innovations into products, and to get to work quickly on superior next-generation products.

Networks of Firms

Japan's large enterprises have developed an impressive organizational solution to avoid the problems of bureaucratic inertia and managerial sclerosis that plague our large Fordist corporations. They cultivate well-organized networks of outside suppliers to keep corporate size down while still gaining the advantages of scale. Ken-ichi Imai, one of Japan's leading industrial organization specialists, reports that 90 percent of the parts used in Fuji-Xerox products, 70 percent of the parts used in NEC and Epson products, and 65 percent of the parts used in Canon products are actually made by outside suppliers and subcontractors.[23] Hitachi, for example, has an immediate industrial galaxy of more than seven hundred suppliers, many of which it partly owns.

Large companies are the linchpins of this new network system of production. They act as anchors, or hubs, for the entire network, which extends outward in concentric rings of first-tier, second-tier, and third-tier suppliers. Suppliers often locate in close proximity to the hub company so that they can deliver components as they are needed, a basic requirement of Japan's "just in time" system of industrial production. There is frequent communication and interaction among the various members of the production system, especially between hub companies and their first-tier suppliers. Frequently, suppliers work alongside hub companies in the development of new products.

The network itself is supported by strong economic imperatives. Suppliers are afforded long-term contracts and are treated well because they are a crucial part of a firm's production system, providing high-quality, reliable parts and on-time deliveries. They in turn work hard to meet demands, since their customers can either move production back in-house or find another supplier. The network takes its shape from deep social relationships—much more so than from simple market-based transactions. The British industrial sociologist Ronald Dore refers to the Japanese system as one of "obligational subcontracting."[24] This highly stable system differs markedly from the constantly shifting, arm's-length supplier relationships characteristic of Silicon Valley and Route 128 (discussed in chapter 6).[25]

And the network creates its own advantages. First and foremost, it allows central hub companies to stay lean and avoid large bureaucracies dedicated to coordinating every little activity in the entire production chain. In doing so, it creates a huge outside source of product and manufacturing process improvements, as suppliers and hub companies work together to make new innovations. These networks become conduits for a rapid and continuous flow of information. The importance of this information exchange is reflected in the fact that at Honda's new R&D lab at Tochigi, for example, adjacent land is set aside for suppliers to build their own R&D facilities.[26] This collaborative aspect shifts the risk of innovation from individual firms to a network of firms, making innovation more attractive and more feasible. When the system operates properly, it can generate powerful technological synergies between companies. This combines the benefits of scale with the advantages of smaller size and flat management hierarchies into a new form of corporate organization that Masahiko Aoki refers to as "quasi-disintegration."[27]

The network also provides a unique vehicle for turning innovations into new companies through "sponsored spin-offs." A classic example of this is Fanuc, a Fujitsu spin-off, which currently ranks among the world's leading producers of industrial robots.[28] Sponsored spin-offs are based on the concept of "growth through connection." Spin-off companies begin life within the corporate parent until they are large enough to leave. The parent provides financing, retains significant ownership, and ensures a permanent relationship with the new company. For example, when Nippon Steel recently formed its rapidly growing software functions into a spin-off business, the new company was guaranteed all of Nippon Steel's information-processing business and any other business the parent could drum up. Interestingly, many of Japan's leading electronics hardware companies, such as NEC, Toshiba, and Fujitsu, are also spinning off their software operations, because these firms are hardware-oriented and find they cannot adequately manage software employees.[29] As time progresses, the spin-off is gradually weaned until it becomes a free-standing member of the parent's industrial network.

Spin-offs can grow large enough to seek out new business on their own, loosening their ties to the original parent.[30] Nippon Denso, a Toyota spin-off, has grown into Japan's leading manufacturer of automotive lighting and electrical systems and in the process has become a major supplier to other Japanese automobile companies.[31] And a few spin-offs ultimately grow larger than their parents. Today, for example, Fujitsu is significantly larger than its parent, Fuji Electric. Such growth is itself beneficial, since it allows the fruits of innovation to diffuse through the economy as new connections are formed with other firms.

Sponsored spin-offs allow large Japanese corporations to move into new areas unhampered by bureaucracy and preconceived no-tions. And they enable them to do so while avoiding both the costs and the risks of the U.S. breakthrough approach.

Product Development: The Cornerstone of Follow-through

Japan's powerful follow-through capacity is bolstered by a heavy emphasis on product development, the D in R&D.[32] Japan devotes nearly nine-tenths of its total R&D effort to product development.[33] And a large majority of Japanese researchers are actually product development engineers—a striking contrast to the U.S., where the scientist with a Ph.D. is far more common. Japan, for example, has only about one-tenth the Ph.D.'s working in the field of biotechnology than does the U.S.[34] In fact, much of what Japanese corporations call basic research is really very advanced applied research. According to a recent study:

> Japanese and U.S. researchers define basic research somewhat differently. Japanese basic research tends to be more goal-oriented; U.S. researchers would probably call it advanced development. . . . Even basic research, therefore, must usually be aimed at a well-defined market goal and have a high probability of success. In the United States, in contrast, basic research is often undirected, with no particular product goal in mind.[35]

Japan's product development focus is not distracted by a heavy military orientation. More than two-thirds of Japanese R&D is funded by industry—giving it the highest share of industry-financed R&D among the advanced industrialized nations (see figure 8.1). In fact, Japan focuses nearly all of its R&D on commercial products and technologies. While Japan spends roughly the same share of GNP on R&D as the U.S., the share of GNP invested in commercial R&D is a full percentage point higher (see figure 8.2).

Masaru Ibuka, honorary chairman of Sony, points out the advantages of Japan's focus on civilian rather than military technology.

Figure 8.1   Percentage of R&D Financed by Industry

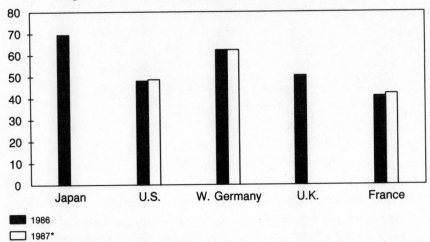

SOURCE: National Science Foundation, *International Science and Technology Data Update 1988* (Washington, D.C., December 1988).
NOTE: 1987 figures unavailable for Japan and the U.K.

Figure 8.2   Spending for Nondefense R&D (as percentage of GNP)

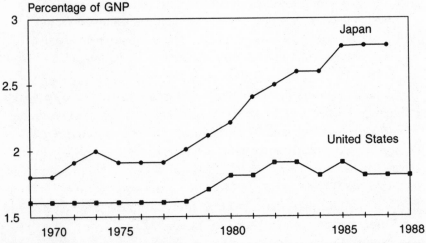

SOURCE: National Science Foundation, *International Science and Technology Data Update 1988* (Washington, D.C., December 1988).

Many advanced technologies were developed in Western nations after World War II. But most of those technologies were aimed at specific military or space development purposes, or for specific manufactured products. There were very few people or companies that thought of utilizing them for the masses. Japanese have been good at applying the technologies to civilian uses. They have all competed to supply the best technologies to the masses. These efforts have allowed the technologies to establish industries and has led to the present prosperity.[36]

Japanese corporations organize research and product development in teams of between five and fifteen scientists and engineers. These teams are typically "self-organizing"; that is, members volunteer rather than being assigned, and they typically form the core of larger, multifunctional teams that will take new products from the idea stage through actual production. R&D scientists and engineers are also encouraged to use company time and equipment to work on their own ideas or "unofficial projects." This allows individual scientists and small groups to take initiative and begin work on new ideas without having to face the scrutiny of management review. If they prove successful, these unofficial projects can slowly turn into official projects with real budgets, larger staffs, and management backing; if not, they can be scuttled at very little cost.

Some Japanese companies are trying to create highly interactive, high-motivation "hothouse" environments found in Silicon Valley and Route 128 start-ups. The development of Sony's new engineering workstation is a case in point. When Sony decided to develop this new product, it organized a team of eleven scientists and set them off in their own space with instructions to develop a new "dream machine" within a year.[37] The team developed a high level of internal motivation and members worked extraordinarily long hours to achieve their goal. As the project moved along, the team even set up mattresses so members could sleep in the lab. Working in this high-motivation, high-effort environment, the team took less than one year to develop a working prototype; and Sony was able to start producing workstations barely two years after the project's inception. As this example illustrates, Japanese corporations are able to effectively replicate both the high-motivation environments and the rapid development times of U.S. start-ups. And, since employees stay with the company for their entire career, Japanese firms can do so without incurring the high costs of disruption and turnover caused by hypermobile U.S. think-workers. In effect, Japanese corporations are able to use their internal labor markets to perform a function that is similar to the external labor markets of Silicon Valley and Route 128.

The organization and orientation of Japan's approach to product development has deep historical roots. Japan's early efforts, like those of the U.S. more than a century ago, focused on "reverse engineering" and improving foreign technologies. Though Japanese companies had always purchased some foreign technology, it was only after the liberalization of technology imports in the mid-1960s that Japanese companies began importing technology on a massive scale.[38] American companies, seeking easy returns on R&D investments and displaying a naïve technological chauvinism, aided and abetted this by granting licenses for everything from televisions to chemicals.

But minor improvements were just the first step. Throughout the postwar years, Japanese corporations were honing their ability to improve foreign products and to innovate in their own right. In 1953, for example, Sony bought a license from the Bell System to produce transistors. Barely two years later and only six months after Texas Instruments, Sony had developed a workable transistor radio—quite a shock to U.S. firms. A similar process occurred later with television and video cassette recorder technology. It is now a familiar refrain in high-technology fields.

Harvey Brooks provides vital insight into the role of imitation (or what we prefer to think of as reverse engineering and improving) in Japan's later success at innovation:

> Although Japan's success in recent years has been based on the adaptation of Western, mainly American technology, and the capacity to commercialize it more rapidly than its competitors, it would be very wrong to conclude from this that the Japanese are mere imitators who, once they have attained the world state of the art in a field, will not continue to move forward the frontier of technology. . . . Successful imitation far from being symptomatic

of lack of originality, as used to be thought, is the first step in learning to be creative.[39]

In his fascinating account of the development of computers and communications in Japan, former NEC chairman Koji Kobayashi provides a stunning list of important innovations pioneered by Japanese corporations during the 1950s and 1960s independently of the U.S. or Western European corporations. These run the gamut from telecommunications switches and cables to new optical fibers and communications satellites. According to Kobayashi, NEC began to develop its own semiconductors in the early 1950s and to work on personal computers by the late 1950s.[40] These early projects and innovations created a solid technology base in Japan upon which later successes would be built.

In the past decade large Japanese corporations have become more expert at improvement, better at product development, and increasingly more innovative.

A New R&D-Manufacturing Synthesis

A key factor in Japan's impressive product development capability lies in the close connection of R&D and manufacturing.[41] Part of the connection stems from the fact that Japanese R&D labs are located on or near actual factory sites; the rest comes from organizational overlap and integration between R&D, product development, pilot production, and manufacturing. Japanese commentators use the term "sashimi system" to describe this system—a reference to a Japanese dish in which raw fish is arranged on a plate in overlapping slices. Hirotaka Takeuchi and Ikujiro Nonaka, leading students of Japanese management, contrast Japan's "rugby" approach of product development, in which the ball gets passed among the team members as they race up the field, to the more traditional "relay race" approach found in the United States.

Under the [U.S.] approach, a product development process moved like a relay race, with one group passing the baton to the

next group. . . . Under the rugby approach, the product development process emerges from the constant interplay of a . . . team
whose members work together from start to finish. Rather than
moving along in defined, highly structured stages, the process is
borne out of the team members' interplay.[42]

No matter what it is called, the key to the process is overlap and
integration of activities.[43] This approach results in a powerful process
of "functional integration"—which stands in sharp contrast to the
U.S. system of functional specialization.

Basically the process works like this. At the initiation of a new
product development project, a team of R&D specialists, product
development experts, manufacturing engineers, and marketing types
is formed. During the early stages, engineers from manufacturing
divisions will come to the R&D labs to prepare for the transfer of a
product to manufacturing. As the project progresses, members join or
depart, depending upon the skills and expertise needed at particular
stages. R&D scientists drop off and are replaced by more members
involved in the later stages (i.e., product development and manufacturing experts). The few senior managers who accompany projects
through the production stage function as "carriers" of projects and
often assume permanent posts in manufacturing facilities. At times,
the entire team moves from an R&D center to a manufacturing site
for initial prototyping and eventual implementation. At an early stage,
representatives of key suppliers also join the team. Although team
members change as the project moves along, the same team will work
on the project from initiation until early production runs are completed.

This system is enhanced by a number of elements that are typical
of the Japanese system. Japanese workers tend to stay with one company for life, which means that all information, ideas, human resource
investments, and learning are retained by the company. In fact, a
recent study found that nearly three-quarters of Japanese engineers
had only one employer during their entire career, a far cry from the
U.S. pattern of hypermobility described in chapter 5.[44] At NEC, for
example, approximately 50 percent of R&D scientists move to operating divisions during their first ten years with the company; after

twenty years about 80 percent have done so.[45] Since many young R&D workers live in company dormitories for the first ten years of their career, they can simply move to a new dormitory at a different site.[46]

Constant interaction between employees with different skills such as marketing personnel, electrical engineers, and manufacturing engineers generates important benefits. Having hands-on personnel involved at early stages ensures that scientists and engineers do not come up with ideas that are too difficult to implement. Instead of the "not invented here" syndrome that often stymies American efforts, the Japanese have a powerful system of collaborative problem solving and organizational learning.[47] And of course, such interaction makes it easier to turn innovations into products and develop new generations of a product.

An important consequence is that products are designed in a way that makes them easy to manufacture. A good example of "design for manufacturability" can be seen in the case of the Plus Hard Card—a computer hard disk on a card produced as a joint Japanese-American venture. The original American design, developed in isolation from manufacturing, was poorly prepared for actual factory production. When Japanese engineers got this design, they used their detailed knowledge of manufacturing to redesign it so that it would be easy to produce.

> Frequently the [Japanese] engineers called for design changes to enhance manufacturability, even if there was not a great cost savings. For example the actuator latch on the disk drive was redesigned so that it would lie against the metal curve of the disk inset in its relaxed position, allowing the disk to be easily inserted (or taken out for service). The new design saved . . . approximately one penny in labor cost per unit, but it made the product more "manufacturable."[48]

And the collective nature of the process generates additional sources of creativity and innovation. Taiyu Kobayashi, a former president of Fujitsu, highlights these advantages:

I believe the strength of Fujitsu lies in our group approach to research. . . . From what I have heard, it appears that individual abilities are given extremely high evaluation in the U.S. . . . I am often told by my friends in competing companies, "You don't seem to have anybody with talent, but you sure get the job done!" Sort of a welcome insult, I suppose. . . . We place more value on cooperative development in which everyone has a sense of participation.[49]

Japan's unique R&D-manufacturing synthesis creates a powerful interplay between innovation and production, reducing the time it takes to turn ideas into actual products. And it enables Japanese corporations to produce a constant stream of innovations—in sharp contrast to the discrete jumps of the breakthrough economy.

Manufacturing and Follow-through

Japanese companies have a tremendous commitment to manufacturing, one that extends to the core of the manufacturing process: the role and responsibility of shop-floor workers.

There is currently some debate over the reasons for Japan's phenomenal success in manufacturing. Some American critics believe that the main reason for this lies in the ability of large Japanese corporations to pump maximum work effort out of shop-floor workers by increasing the pace of work, forcing employees to work long hours, and using teams to keep the pressure on workers.[50] There is no doubt that Japanese workers work very hard: they work more hours than their American counterparts—on average 41.3 hours versus 37.3 hours per week. Workers in supplier firms receive lower wages and face more onerous working conditions than the more privileged strata of employees in large companies. And there is a large stratum of temporary workers, typically women, who toil in sweatshop conditions. Japanese firms have been very effective in eliminating idle downtime and waste—at filling in the pores of the working day.

But "hard work" alone cannot explain Japan's remarkable success in industries ranging from steel and automobiles to microelectronics and biotechnology. If hard work alone were the key, Korea—not Japan—would be the world's leading high-technology producer. The key to Japan's success lies, rather, in a new form of work organization that taps the brains as well as the brawn of shop-floor workers to create a synthesis of smart work and hard work.[51] This new form of tapping intelligence as well as physical labor harnesses more completely, more totally than under Fordism. Under this system, all workers become "knowledge-workers." The words of the late Konosuke Matsushita, founder of the Japanese electronics company that bears his name, capture the essence of this synthesis.

> We are going to win and the industrial west is going to lose out; there's not much you can do about it because the reasons for your failure are within yourselves. Your firms are built on the Taylor model. Even worse so are your heads. With your bosses doing the thinking while the workers wield the screwdrivers. . . . For you the essence of good management is getting the ideas out of the heads of the bosses and into the hands of labor. We are beyond the Taylor model. Business we know is now so complex and difficult, the survival of firms so hazardous and fraught with danger, that continued existence depends upon the day-to-day mobilization of every ounce of intelligence.[52]

Matsushita's words provide keen insight into the kernel of the Japanese system: its ability to tap the intelligence of workers at all levels.

This new synthesis is itself a powerful component of follow-through capacity, providing an unparalleled source of production advantages and an important source of product and process innovation in its own right. In recent studies, Haruo Shimada advances the concept of "humanware" to describe Japan's smart production workers, while Kazuo Koike uses the idea of "learning by doing" to convey the way Japanese firms use intelligence to improve production. The advantages of this new system are especially evident in microelectronics, where the combination of hard and smart work help to optimize the capabilities of the entire production systems. James Koford characterizes the advantages of the Japanese system: "I'm very frightened of the Japanese. Japanese semiconductor facilities have state-of-the-art equipment. They have the best manufacturing equipment. And there are thousands of workers using that equipment and busily tweaking aspects of everything to make the whole process better, more efficient."[53]

The synthesis of hard and smart work is an important element in Japan's rapid move to the highly automated "factory of the future." Japan's smart workers have the capabilities needed to effectively operate automated manufacturing equipment and do not fear that automation will displace them. They function as the human links between various islands of automated technologies and a buffer against machine failure.[54] The combination of knowledge-intensive workers and new flexible manufacturing technologies is shaping a second industrial revolution in Japan that is changing the face of mass production as profoundly as the Fordist assembly line. For example, Japanese automobile factories can now produce different models and variations of cars one-by-one on the same assembly line.

The tremendous power of the Japanese manufacturing system is becoming increasingly obvious to Americans, as more and more Japanese companies set up factories in the United States. Our own research shows that Japanese corporations are successfully "transplanting" this system in industries ranging from automobiles, rubber, and steel to computers, semiconductors, and other high-tech products. The Honda assembly complex in Marysville, Ohio, for example, essentially uses the same technology and production organization found in Honda's Japanese plants.[55]

The organizational centerpiece of the Japanese system is the "self-managing" work team.[56] Teams are composed of between three and fifteen workers and are responsible for the organization and distribution of work to team members, basic quality control, and correction of problems that crop up on the line.[57] At Kyocera, workers are organized in "amoebas," which are self-managing teams that are formed and dissolved according to project needs. Teams serve a series of complementary functions. They are a basic mechanism for generating internal, self-imposed discipline, devolving managerial responsibilities to the shop floor, and motivating workers to work harder.

They are also a mechanism for harnessing workers' knowledge and collective problem-solving capabilities for the enterprise.

Other factors contribute to close interconnections between shop-floor workers, engineers, and managers. The concept of "management by walking around" takes on new meaning in Japanese factories. Engineers and managers are housed in offices that are adjacent to the shop floor, and they spend a considerable amount of time on the factory floor, talking to shop-floor workers and devising on-the-spot solutions to problems. At Honda's assembly plant in Marysville, there are not enough desks for all the junior engineers. They are thus forced to walk around the plant and stay in touch with what is happening on the shop floor.[58] In addition, most managers and engineers spend the first decade or so of their careers as lower-level supervisors (often working even in factory jobs) and are members of the company union, so they develop a better feel for actual manufacturing activity and a greater appreciation for the practical knowledge of shop-floor workers.[59]

The use of consensus decision making builds in additional levels of communication and interaction, guaranteeing that all, or at least most, employees buy into new projects. This process also maximizes the flow of information and knowledge throughout the company. Kao Corporation has a policy of having all of its computers open to employees (with the exception of personnel files), the objective being to encourage employees to learn about the company and contribute their ideas.[60]

Japan's advances in manufacturing have deep historical roots. The period immediately after World War II was one of intense labor-management conflict. In some cases, workers even took over factories and implemented radical "production control" strategies, running the plants without management.[61] It was these struggles that transformed Japanese industrial relations and created the institutional framework upon which recent Japanese advances in the organization of manufacturing are built.

Basically these struggles set the stage for a new social compact, or accord, between Japanese business, labor, and government. This accord established the institutional framework for Japanese industrial relations.[62] At the heart of this accord was a critical trade-off. Japanese blue-collar workers won the right to be considered as part of the core employees of the firm, thereby receiving an implicit guarantee of employment.[63] In return for this, workers lost their original demands for control over the organization of production and the right to specific jobs. Thus, management won great leeway over the organization of work.[64]

The accord brought about a major rethinking of the role and function of manufacturing workers. On the one hand, long-term employment made workers a fixed cost; on the other hand, it enabled the firm to capture all of its investments in training or skill improvement. Eventually, management began to see workers as an asset that could contribute to improved quality, increased productivity, and innovations in shop-floor product and process. As this view took hold, Japanese corporations endeavored to erect new barriers to mobility, such as generally not hiring away each other's workers. In this way they protected their investment in their labor force.[65] In addition, the tenure system placed tremendous pressure upon management to develop new products and technologies that could absorb labor. The result was a powerful new organizational system that was a major advance over the artificial separation of "hands" and "brains" found in American corporations, large and small alike.

Still the benefits of the Japan's manufacturing system accrue mainly to male, permanent employees in large companies, who make up just one-third of the work force. Surrounding this "core" of the labor force is a periphery of lower-paid, part-time, and temporary workers, especially women, who work for subcontractors, suppliers, and small firms and who do not benefit from the conditions afforded core workers.[66] Japan continues to suffer from serious problems of racism, sexism, conformism, overwork, and exploitation. Of course many aspects of workers' standards of living, especially housing, are lower than those in America. And it should not be forgotten that Japanese corporations harness the intellectual and physical capabilities of all their workers more completely and more effectively than other systems. Simply put, the Japanese are better capitalists, but they are not angels.[67]

Adding Breakthrough to Follow-through

One fundamental question remains: Can Japan add breakthrough capacity to its already well-honed follow-through capability?

While most Americans continue to see Japan as merely a follower and purchaser of "superior" U.S. technology, a growing group of experts believe that Japan is becoming more innovative. James Abegglen and George Stalk suggest that "the kaisha (Japanese industrial firms) as competitors are driving for technological leadership. Western competitors were caught by surprise when the kaisha achieved cost and quality levels fully competitive in many industries. Will Western competitors be surprised again as technological parity, and leadership, is achieved?"[68] Thomas Murrin, a former Westinghouse executive and deputy secretary of commerce believes that "the Japanese are ending a period in which most of the technology they exploited came from the U.S., and to a lesser extent, from Europe. But Japan, just like the U.S. before it, is increasing its R&D investments and tremendously improving its ability to do laboratory R&D."[69] Japan is no longer a mere follower of U.S. technology but a powerful innovator in its own right.

This is illustrated by the growing number and quality of U.S. patents granted to Japanese inventors. Japanese companies increased their share of U.S. patents from 4 percent in the 1960s to more than 20 percent by 1989.[70] Today three Japanese companies—Hitachi, Toshiba, and Canon—top the list of U.S. patent getters, ahead of IBM, DuPont, General Electric, Westinghouse, and many others (see table 8.2).[71] Japanese companies now account for more new U.S. patents than do the United Kingdom, France, and West Germany combined. Further, these patents are concentrated in "hot" commercial areas such as semiconductors, scientific instruments, consumer electronics, and, of course, automobiles; and they are among the "most highly cited patents," that is, the most commercially relevant. On the basis of these facts, a recent study concludes that "the Japanese position in patented technology is very strong, growing steadily, and based on a high-quality, high-impact technology that has been invented by Japanese inventors."[72]

TABLE 8.2

Companies Receiving Largest Number of U.S. Patents in 1988

Rank	Company	Country of Origin	No. of U.S. Patents
(1)	Hitachi	Japan	907
(2)	Toshiba	Japan	750
(3)	Canon	Japan	723
(4)	General Electric	United States	690
(5)	Fuji Photo Film	Japan	589
(6)	Philips	The Netherlands	581
(7)	Siemens	West Germany	562
(8)	IBM	United States	549
(9)	Mitsubishi Denki	Japan	543
(10)	Bayer	Germany	442

SOURCE: "A Foreign Push in U.S. Patents," *New York Times,* June 4, 1989.

Japanese firms are massively increasing R&D spending as they endeavor to reorient themselves from commodity production toward knowledge-intensive products. As figure 8.3 shows, Japan's R&D expenditures have skyrocketed over the past two decades, more than tripling in real dollars. And since 1965 the proportion of the Japanese labor force engaged in R&D also tripled, increasing from 2.5 percent to 6.2 percent of the work force. Between 1982 and 1987, R&D spending by Japanese corporations grew by 60 percent. In the past few

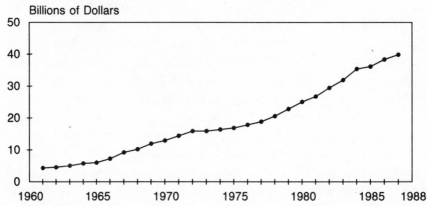

Figure 8.3   Japanese R&D Expenditures, 1961–88

SOURCE: National Science Foundation, *International Science and Technology Data Update 1988* (Washington, D.C., December 1988).

162

years, R&D spending has grown at an annual rate of 15 percent. The magnitude of this investment is illustrated by the fact that in 1989 the top ten Japanese electronics firms invested more in R&D than the top one hundred U.S. firms.[73]

Top Japanese executives indicate that R&D expenditures and personnel must increase in the future. This is true both in high-technology industries, where R&D spending is seen as a way to move ahead of U.S. and European competition, and in mature industries, like steel and textiles, where R&D is a needed to move into new high-growth areas.[74]

Japanese companies are working on other fronts as well. Many have placed R&D scientists in American labs and universities in an attempt to learn more about our breakthrough system. A large number are engaged in joint research projects with both large and small U.S. companies. Some are opening R&D labs in the U.S., especially in Silicon Valley, to learn more about the U.S. system of breakthrough innovation.[75] Others are investing in American venture capital funds in order to gain accelerated access to new breakthroughs.[76] And many are investing in American universities to endow professorships, join industrial liaison programs, and build new laboratories on American university campuses.[77]

As comforting as it may be, it is a serious mistake to believe that Japan is simply trying to duplicate our model of breakthrough innovation. Japanese companies are actively adding a stronger R&D capacity—a breakthrough capacity—to their already well developed follow-through structure.[78] Japanese companies are making massive increases in capital investment so that they can turn new innovations into mass-produced commercial products. According to an April 1990 report in the *New York Times,* in 1989 Japanese capital investment hit $750 billion, or 24 percent of the gross national product. The U.S., which is a much bigger economy, invested just $500 billion, or 10 percent of the gross national product.[79]

Japan has clearly moved beyond the organizational rigidities and problems that plague both our old Fordist firms and the high-technology start-ups of the breakthrough economy. For us, the Japanese system represents a qualitative break with both, an alternative, if ideal-typical, model of technological and industrial organization that we call "Fujitsuism." Fujitsuism is adapted from the name of one of Japan's largest and most important high technology companies, Fujitsu Ltd., which recently replaced IBM-Japan as Japan's largest computer maker. Its spin-off Fanuc is currently the largest robot manufacturer in the world—at the Fanuc factory in Japan untended robots work in the dark producing parts of other robots. Fujitsuism is distinguished by the synthesis of intelligence and hard work, which is accomplished through high levels of functional integration on the shop floor, among networks of firms, and between production and innovation. The consequences are indeed powerful: the synthesis of hard and smart work dramatically enhances the potential for innovations to come from the shop floor, while integration across the R&D-manufacturing spectrum creates an environment in which innovations of all sorts are fostered and rapidly translated into products with real economic impact.[80]

Many in America are finally recognizing the depth and seriousness of the competitive threat posed by Japan's remarkable organizational advances and follow-through capacity. Both industry and government are starting to develop strategies to bolster American competitiveness. It remains to be seen just how effective these strategies can be.

9
Patchwork Solutions

■

In the United States we have no sense of community. Instead we have a nation of self-interest. Get mine. Let the other guy worry about getting his. No shared dream of how we can make this a better place. . . . There is no "America Tomorrow" plan. Instead, we all grab for what we can and hope tomorrow will take care of itself.
—FRANK BURGE, publisher of *Electronic Business*[1]

Industry and government are responding to our technological decline with new programs designed to restore America's edge. These run the gamut from corporate restructuring, factory automation, and the reorganization of R&D labs to takeovers, strategic alliances, R&D consortia, and new forms of intellectual property protection. Large corporations are moving forward on a number of fronts. In her new book, Rosabeth Kanter outlines how they have begun to eliminate management layers, reduce hierarchy, form new partnerships and alliances with outside organizations, and implement new remuneration schemes that emphasize pay for performance in an effort to make themselves more competitive.[2] Recent reports indicate that about one-quarter to one-third of all American companies and more than half of all large companies have instituted new schemes designed to increase employee participation.[3] Motorola has set up work teams, a bonus system that rewards workers for increases in productivity, and a new education and training program to improve employees' basic mathematics skills used in worker quality control.[4] Xerox has instituted employee participation programs in seven plants.[5] GE has established work teams, a new "pay for knowledge" system that rewards workers for obtaining new skills, and a company-financed training program.[6] But this is only a start. Our leading corporations will need to go much further to ready themselves for the challenges of the twenty-first century.

Corporate America's attempts at reform are on such a limited scale that they amount to little more than superficial restructuring. In fact, restructuring is frequently used to mask corporate attempts at downsizing—a strategy for elimination of jobs through increased work effort, automation, and use of subcontractors.[7] Under this strategy, work reforms are simply an excuse to streamline operations, preparing the way for them to be contracted out or moved offshore. In numerous cases, changes in corporate strategy have led to the summary closing of reorganized plants, with little regard for displaced workers. Xerox, for example, laid off a large number of workers less than two years after it instituted work reorganization. GM's decision to close its much-heralded Fiero plant is another case in point.

Restructuring is more hype than reality in many U.S. firms. Work teams may be introduced, but they are not combined with other reforms that are needed to make them successful. Few companies have combined teams with decentralized decision making. Absent this, teams are simply a new mechanism of grouping workers that will not produce meaningful productivity improvements.[8] Too often, quality control circles are simply off-line "discussion groups," not an integral component of the actual manufacturing process.[9] Given this low level of commitment, it is little wonder that more than 20 percent of experiments in quality control circles fail in their first year. And few companies provide the training that is needed to make work teams function more effectively. According to employee involvement consultant W. Patrick Dolan: "Training is always a peripheral, secondary consideration, and when push comes to shove it keeps sliding."[10]

But the biggest failing of current corporate efforts lies in the continued neglect for shop-floor workers. Wedded to the old Fordist belief that workers are menials, top management has little understanding or respect for their intelligence. Given this attitude, there is little possibility of truly engaging workers in corporate restructuring. Until U.S. corporations recognize the importance of "smart work" to the enterprise, attempts to improve our manufacturing performance will have marginal success.

Corporate America's version of restructuring faces considerable opposition from shop-floor workers. Some workers would rather quit than participate in top-down restructuring. Workers at GE's Salisbury plant, for example, bolted when reorganization was implemented, and turnover rose to 14 percent the first year. Others become demoralized and reduce their work effort. Apparently, some workers prefer to do one boring job, where they can forget what they are doing, rather than several boring jobs.[11] Heated unrest has in fact flared at some supposedly "reformed" factories like GM's Van Nuys plant.

Corporate restructuring is increasingly criticized by trade unionists, academics, and workers, many of whom are understandably convinced that reorganization is just a facade for work intensification and the elimination of jobs. John Brodie, president of the United Paperworkers Local 448 in Chester, Pennsylvania, sums up the skepticism of many labor activists: "What the company wants is for us to work like the Japanese. Everybody go out and do jumping jacks and kiss each other and then go home at night. You work as a team, rat on each other, and lose control of your destiny. This is not going to work in this country."[12]

Brodie's sentiments are understandable, for this is the reality of the "team concept" as practiced by many American corporations. Like many others, he harks back to the "golden age" of American industry and industrial unions, that is, to our old follow-through economy. Brodie's words reflect the adherence of the trade union movement to the outmoded rules and organizations of mass production industry— structures that have already proven quite vulnerable to both foreign and domestic competition. His assumption that workers in our industrial corporations once had greater "control of their own destiny" is dead wrong. Workers in American industrial corporations were little more than cogs in a machine, not permitted to think or use their intelligence in their work. Armed with such an outmoded philosophy, today's industrial workers are in charge of their destiny, only insofar as their destiny is to be unemployed. Unions too need to adapt their strategies to the new realities of the high-tech age.

The restructuring of the corporation currently touted by management consultants seems more geared to shifting cost to lower-level employees and shielding top executives from the implications of weakening industrial capabilities. True restructuring will require a fundamental shift in the power relationship between stockholders, management, and workers—away from the first two and toward the workers.

The "Factory of the Future"

At the same time that they endeavor to restructure themselves, large corporations are also trying to automate their problems away by eliminating labor from direct production.[13]

A host of Fortune 500 firms—IBM, GE, Westinghouse, GM, and Texas Instruments, to name a few—are reequipping their production lines with high-tech automation. According to Dataquest, purchases of factory automation systems increased from less than $10 billion in 1980 to over $18 billion by 1985.[14] Currently, there are some thirty fully automated plants and thousands of partially automated facilities in the United States.[15]

The U.S. approach to factory automation is conditioned by the contempt American corporate leaders have for production workers. Most companies are using factory automation to create an automated version of the Fordist factories of the old follow-through economy. In an important paper on the subject, Ramchandran Jaikumar suggests that U.S. managers treat automated manufacturing

> as if it were just another set of machines for high-volume, standard-ized production—which is precisely what it is not. Captive to old-fashioned taylorism and its principle of scientific management, these executives separated the establishment of procedures from their execution, replaced skilled blue-collar machinists with trained operators, and emphasized machine uptime and productivity. In short, they mastered narrow-purpose production on expensive FMS technology designed for high-powered, flexible usage. . . . Certainly, Frederick W. Taylor's work still applies—but not to

this environment. Managing an FMS as if it were the old Ford plant at River Rouge is worse than wrong; it is paralyzing.[16]

American companies fail to see the importance of smart workers to factory automation. Operating with outmoded conceptions, they see automation mainly as a way to get rid of workers, remove responsibility from the shop floor, and transfer workers' practical knowledge to management.[17] For these reasons, automation is often a source of labor strife. A highly motivated and skilled work force can become a major asset when automated equipment is inoperative or needs "tweaking" to make it operate effectively. Lack of attention to workers' roles or training is a fatal mistake.

American companies have often been more concerned with getting systems up and running rather than with anticipating problems and building in reliability. Even minor glitches can cause highly automated systems to crash. For example, an incorrectly placed ink spot caused problems at IBM's ProPrinter plant for six months.[18] Software problems have caused tremendous delays at GM's automated plants. According to Peter Hartwell, director of manufacturing for GM-Canada, "EDS wrote all the software. But then it didn't work. . . . The bottom line was that rather than the systems making us more productive, they were slowing us down."[19] And of course, the workers just stood aside and laughed: far be it from them to suggest what the problems might be.

Factory automation is heavily dependent upon Defense Department subsidies and military users. A 1989 *Business Week* article waxed eloquent over the improvements in U.S. manufacturing prowess, but of the twelve firms mentioned seven were heavy defense contractors.[20] Recent articles in the business press have been effusive in their praise of the automated manufacturing techniques that have been used in the Stealth bomber program. What was neglected was that these bombers will certainly be the most costly ever built. The *Wall Street Journal* recently suggested the plane's price tag might exceed its hundred-ton weight in gold.[21]

Large U.S. companies see automation as a cure-all for a host of other organizational problems as well. Many companies want to use automated systems to streamline their international purchasing patterns, in effect, to create international just-in-time systems. But there is no way that computerized communication can compensate for long distances and the lack of physical proximity.

Corporations have poured countless billions into office automation in an attempt to rationalize back-office and clerical functions. Yet the "office of the future" has been an even bigger disaster than the factory of the future. According to data compiled by Stephen Roach of Morgan Stanley, while factory automation has generated some small productivity increases, the productivity payback from office automation has been negligible.[22] This should not be surprising once we realize that American companies have poured capital into new office technology in an attempt to avoid the real organizational restructuring they need.

Technological "fixes" like factory or office automation cannot compensate for organizational weaknesses.[23] Unable to divorce themselves from the breakthrough ideology of this economy, American industrial leaders keep hoping that the next breakthrough will be sufficient to leapfrog them ahead of their competitors and save the day. Unfortunately, this will not work. In today's global marketplace, virtually anyone can buy the automated machines they need to compete. Factory automation will be a success only when combined with sweeping organizational restructuring. The victors in the race to the factory of the future will be those who have the most highly skilled and committed workers.

Reorganizing the R&D Lab

The reorganization of the R&D lab is still another in the long list of solutions that large corporations are trying out. Digital Equipment Corporation (DEC), for example, has set up "product business units" to link R&D, product development, and manufacturing. Borrowing from Japan's multifunctional teams, Codex, a Motorola subsidiary in Canton, Massachusetts, has recently set up core teams that shepherd products from conception through manufacturing and distribution, cutting its development time in half.[24]

While such efforts are a step in the right direction, it is difficult to gauge their depth and seriousness. Are they long-term strategies or merely cost-cutting expedients? Will corporations develop supplementary policies, such as the permanent rotation of personnel from the R&D labs to factories, that can bring R&D and manufacturing closer together? Can the arrogance of the white-collar scientists and engineers toward shop-floor workers be overcome? Can a few successful examples spur a broader effort on the part of American industry? On these counts, we are not so sanguine. Despite a few pioneering efforts, most corporate R&D labs retain the specialized, assembly-line model of organization discussed in chapter 2.[25]

A naïve sense of technological supremacy still pervades our corporate R&D establishment and thus inhibits the kinds of reorganization that are needed to be successful. For example, few U.S. researchers are willing to spend time in Japanese R&D labs (even though slots are open to them), because they believe those labs are not up to the standards of American R&D centers. In 1989 *Business Week* estimated that there were fewer than 150 U.S. scientists working in Japanese labs. Moreover, American scientists who have spent time in Japan point out that their American employers are uninterested in learning about Japan's technological strengths or the way it organizes R&D.[26]

Worse yet, corporations are slashing their R&D budgets as they endeavor to restructure by cutting costs. In 1989, corporate R&D spending failed to keep pace with inflation for the first time in a decade and a half. According to the National Science Foundation, corporate R&D spending rose by just $2.3 billion from $66.5 billion in 1988 to $68.8 billion in 1989, a net change of nine-tenths of one percent controlling for inflation.[27]

The era of corporate raids and hostile takeovers has only exacerbated this problem. Faced with escalating foreign competition, dwindling commercial payoffs from R&D, and pressure by corporate raiders to increase dividends, some companies have proceeded with R&D cutbacks and consolidations. And some have even sold off their R&D labs. Goodyear Tire and Rubber Corporation, for example, slashed R&D spending in its battle with corporate raider James Goldsmith, even though its major Japanese and European competitors were increasing R&D investment to remain competitive. GE recently sold off the world-renowned David Sarnoff Research Laboratory it inherited in its purchase of RCA.[28] According to a recent study by the National Science Foundation, twenty-four companies that were involved in mergers and leveraged buy outs in 1986 and 1987 cut R&D spending on average by 5.3 percent; and the eight companies that were involved in protracted takeover struggles cut R&D spending by more than 13 percent.[29] William Spencer, a Xerox vice-president, provides some perspective: "We've moved from research and development being a corporate asset to where it's what a corporate raider looks for first. They can make significant cuts and get cash flow. I haven't seen a takeover yet where they increased research and development activities."[30] While corporate policies such as these may bolster short-term profits, they will only set back efforts to restore American technological competitiveness and industrial might.

Linking Large and Small

Another highly touted solution for restoring America's technological prowess involves combining the innovative capacity of our small start-ups with the manufacturing and marketing prowess of large industrial corporations through alliances and other forms of partnership. A *Business Week* story in 1989 reflects the hype:

> Giant American companies are going the "corporate partnering" route, too. Small-growth outfits are linking up with American Telephone and Telegraph Co., IBM, and Digital, to name just a few. Entrepreneurs are willing to sell equity to these behemoths— and give up some independence—to a get a higher price than offered by the public markets. Corporate giants also exchange marketing and distribution muscle for access to a new technology.[31]

Some see strategic alliances as a basic element of new postmodern management strategies that will bolster American technological capabilities.[32] Harvey Brooks hails the integration of large and small as "probably the greatest strength in the American picture . . . combining some of the stimulation of competition with the benefits of aggregating technical resources and some measure of strategic planning."[33]

Unfortunately, the reality of strategic alliances falls far short of such idealizations. Competition and conflicting objectives make it hard to build true cooperation. Small companies view alliances as only a stepping-stone, or worse yet a last resort, when they need capital or access to manufacturing capabilities. Small companies are especially fearful of losing their "technical birthright." The objective of a small company is to share as little as possible with its "ally." Venture capitalist Eugene Kleiner puts it quite bluntly: "Strategic partnerships from the entrepreneur's point of view are a sellout."[34]

Large companies want access to new technology and are willing to pay for it. As one venture capitalist aptly observes: "Companies are finding that developing new products is expensive, time-consuming, and risky, and that it is cheaper, faster, and safer to buy a proven capability."[35] But once they get what they need, they have little incentive to continue the partnership. In biotechnology, for example, there has been a recurring pattern of large chemical and pharmaceutical companies abruptly canceling their joint venture agreements with start-ups after they have managed to appropriate the start-ups' technology.[36] In reality, the small firm has little recourse to remedy transgressions by its ally. The normal path of filing suit to defend itself is usually not feasible because the cost of litigation often overwhelms its meager resources. In the words of Silicon Valley venture capitalist Reid Dennis: "When a flea sleeps with an elephant, it's awfully easy to get rolled on."[37] The alliance route is often less a partnership than a distrustful marriage of convenience to be broken at the earliest opportunity.

Mergers are another way of linking large and small. And these are fraught with even greater problems than are alliances. Consider the following disaster. In 1981 General Electric decided to reenter the semiconductor industry and purchased Intersil, a classic Silicon Valley semiconductor start-up. GE's president, Jack Welch, vowed to leave Intersil's start-up environment unaltered, but GE's bureaucracy soon began to assert itself. Intersil executives were subjected to GE's maze of procedures and constant requirements for reports. They were constantly traveling cross-country to report in person to headquarters.

The degree of standardization and pettifogging rules was astounding. According to one former Intersil executive, even presentation overheads and slides had to be done according to corporate standards. In another case, an Intersil executive said that he found GE accountants poring over Intersil's books on the weekend, unbeknownst to most of Intersil's management.[38]

Intersil and GE executives clashed repeatedly and bitterly over an increasing number of issues. GE managers took away Intersil's incentive pay program because it was out of line with GE's established salary structure. When GE insisted that Intersil lay off people in response to a semiconductor industry downturn, key Intersil executives began a mass exodus to start-ups and other companies. GE then sent in managers from headquarters, but Intersil continued its downward spiral. As one Silicon Valley executive so eloquently put it: "Intersil became an empty shell. GE was like the kiss of death."[39] By 1988, after hundreds of millions of dollars in losses, GE sold off all of its commercial semiconductor business to the Harris Corporation.[40] Far from creating new value, this misadventure destroyed a viable company and seeded Silicon Valley with more entrepreneurs who hate big companies.

And this is just one of many cases. The list of failures is indeed long: Xerox and Scientific Data Systems, IBM and Rolm, Exxon and a bevy of electronics start-ups, and Eli Lilly and Hybritech. A few such acquisitions have come full circle with the acquired company orchestrating a leveraged buy out to break away from a stultifying corporate parent. After suffering huge financial losses and revolving-door management under Exxon ownership, Zilog orchestrated a successful leveraged buy out with $35 million dollars in capital from E. M. Warburg Pincus Capital.[41] While the company is now doing much better on its own, such financial machinations can be a waste of capital and human talent.

The result of these "elephant and mouse" takeovers is a familiar refrain of high levels of bureaucracy: frustration, eventual defections, and finally abandonment. It is almost impossible for large corporations to generate the atmosphere that high-tech think-workers need. The closer the large firm gets to the small firm, the more the small firm is suffocated by its environment. In the end, the large firm has usually not achieved its goals while the small firm has been devastated. Thus, the "perfect marriage" of innovative start-ups and the marketing and manufacturing prowess of our corporate giants too often ends up in either a painful divorce or a badly failing union that brings out the worst in both partners.

The Foreign Partner Contradiction

There is a fatal flaw in the logic of those who see linkages between large and small firms as the key to a U.S. comeback in high technology. The flawed logic is, quite simply, that foreign companies can and are playing the same game. For example, Steven Jobs's Next recently sold 16 percent of its shares to the Canon Corporation of Japan for $100 million.[42] Nearly every U.S. semiconductor company has a technology-sharing agreement with a large Japanese partner. Small biotechnology firms are also striking a large and growing number of partnership agreements with foreign companies.[43] In just the three-year period from 1986 through 1988, Japanese corporations invested more than $650 million in 120 alliances with venture-backed start-ups.[44] In the first half of 1989, they invested another $214 million in small U.S. high-technology companies, more than two and one-half times what they invested in the comparable period for 1988.[45] And the surge of foreign investment in and acquisition of U.S. high-technology start-ups extends far beyond Japan. In December 1989, Wyse Technology, a Silicon Valley computer company, was bought by a consortium of Taiwanese investors, Channel International Corporation, for $265 million.[46] In February 1990, Genentech, America's premier biotechnology start-up, was acquired by the Swiss pharmaceutical giant Hoffmann-La Roche.[47]

There is growing concern among high-tech industrialists and makers of public policy that foreign alliances and takeovers "give away" important technologies to our competitors. An article in the *New York Times* suggests that cross-national alliances "have resulted in a largely one-way flow of technology and other critical skills from the United States to foreign nations, especially Japan."[48] *Business Week* has recently reported that the growth in U.S.-Japan alliances is "Silicon Valley's version of *Let's Make a Deal*. In a glitzy high rise in downtown San Jose, Calif., a new Japanese consulting firm is on the prowl. . . . [It] is just one sign of what some might call America's Great Technological Sell-Off—a Japanese buying binge that could siphon off some of the most promising ideas being developed by U.S. start-ups."[49]

But despite all the controversy, it is difficult to block foreign access to our breakthrough technologies. The reason for this is simple. The breakthrough economy produces wave after wave of start-ups that need capital to survive after they have used up their initial venture capital. Ever since the 1987 stock market crash, the initial public stock offering market has been weak, making it even harder for companies to move from venture capital to the public market. Start-up and expansion capital has also been harder to get as U.S. venture capitalists turn increasingly to leveraged buy outs, which comprised roughly 20 percent of all venture investments in 1989.[50] According to one Silicon Valley executive, the scarcity of start-up and expansion capital has led to the following situation: "We are looking at every imaginable alternative short of bank robbery."[51]

Foreign companies and foreign equity investors are an obvious and eager source of capital. Relationships with foreign firms are often preferable because they will not immediately compete in the U.S. market. This gives the small company the time and space it needs to expand. Foreign companies can also help American start-ups tap foreign markets. Sun Microsystems has an overwhelming presence in the Japanese workstation market because of its alliances with Japanese companies.[52] A future variation on this theme is the creation of new offshore venture funds and stock markets designed to tap eager equity investors in Japan, Korea, and Taiwan.[53] Hambrecht and Quist, a leading U.S. venture capital firm, has recently opened a branch office in Taiwan to tap into Taiwanese capital. TA Associates, a major Boston fund, has long had a European affiliate, Advent International, to mobilize foreign capital.[54] A March 1990 survey of 750 leading electronics companies found that foreign capital accounts for 11 percent of total capital for high-tech electronics firms and roughly 20 percent of total capital for semiconductor and computer companies.[55] Simply put: Foreign financiers are increasingly becoming the venture capitalists for the breakthrough economy. As this occurs, access to and ownership of U.S. high technology itself will become increasingly foreign.

Thus, the breakthrough economy has created a "technological commons" into which foreign companies can easily buy access. Before the 1980s most Japanese firms were in a "creative imitation" mode and merely bought licenses and reverse-engineered technology; now, however, they aim to secure access to the breakthroughs necessary to provide the technological seeds for their own follow-through economy.

But the main reason U.S. start-ups find it necessary to form alliances with Japanese companies is because large American companies are unwilling to partner with them. According to James Solomon, president of SDA (now Cadence), a Silicon Valley chip-design company, Japanese companies are far more interested in and willing to adopt the technology his company develops. Tazz Pettibone, whose company was acquired by Kubota in 1987, put the issue in stark perspective: "We were running out of money and our [U.S.] venture capitalists were not prepared to go into deep funding."[56] A recent report to the U.S. Economic Development Administration provides an account of how one small company

> spent six months on the road talking to investors trying to raise six million dollars. Among their more preferred investors were two major U.S. corporations. . . . After six months of failure trying to raise money among major U.S. companies, a major firm from the Orient spent one day touring this company's factory and design facilities and the only question asked was, "How much?" While the company has been much criticized by some for selling 40 percent to an Oriental company, its executives point out that they worked 24 hours a day for six months to obtain a U.S. corporate investor, without success, and needed the cash infusion to survive and to capitalize on its research and development investment.[57]

This is not merely anecdotal. It is, rather, symptomatic of what is occurring across the board in high-tech industry. The reason is simple: Under the incentive structure of the breakthrough economy, profit and financial gain are the ultimate and often exclusive goal. For the venture capitalists and entrepreneurs of the breakthrough economy, foreign capital is still capital, a foreign deal is still a deal.

The impassioned attacks upon foreign companies developing liaisons with our small firms are understandable, but they fail to take cognizance of the causes of these relationships. For the venture capitalist Donald Valentine, this problem is largely of our own making: "Neither private nor public markets here are willing to support new technologies."[58]

R&D Consortia: Safety in Numbers

Many see R&D consortia as a "magic bullet" that will restore our technological lead.

R&D consortia are not a completely new phenomenon. Since the turn of the century, there have been consortia such as the American Iron and Steel Institute and the Edison Electric Institute. A 1956 study of R&D consortia by Battelle Memorial Institute identified 173 R&D joint venture arrangements in the United States, 66 of which operated their own laboratory facilities.[59] Former Sematech president Robert Noyce often argued that R&D consortia are "the best way to beat Japan."[60] In the words of William Norris, founder of Control Data Corporation: "The stage is set for industry initiative to expand R&D cooperation rapidly. Only through joint endeavors of companies can the U.S. meet the challenge of its world leadership in technology, enhance free-market competition, and expand the job opportunities of its citizens and consumer choices."[61]

Two of the best known are the Microelectronics and Computer Technology Corporation (MCC) and Sematech. MCC was formed in 1982 by leading microelectronics companies in response to the Japanese Fifth Generation Computer Project.[62] Sematech was formed in 1987 by many of the same microelectronics companies to improve U.S. semiconductor manufacturing capability.[63] Sematech's membership includes IBM, AT&T, DEC, Hewlett-Packard, Intel, and most

178

of the other important semiconductor companies; its funding is divided between the participant companies and the Department of Defense. The federal government has already pumped more than $200 million into Sematech, and industry officials want it to contribute another $800 million over the next three years. T. J. Rodgers, the founder of Cypress Semiconductor Corporation, has remarked sarcastically: "The old semiconductor companies have become very good lobbyists."[64]

R&D consortia face problems of their own. Building the close cooperation it takes to make R&D consortia successful is quite difficult in the highly fragmented environment of the breakthrough economy. A recent report summarizes this problem very well: "Can American high tech companies, which thrive on cut throat competition, learn the seemingly schizophrenic task of working together while still honing their competitive edge."[65]

Companies find it especially difficult to share in areas where they have proprietary interests. This is exacerbated when there is a wide disparity in size, age, and stature of the companies involved. Reconciling the interests of competing firms can be a managerial nightmare. Already at MCC, there have been difficulties in convincing companies like DEC, Hewlett-Packard, and Martin Marietta to share ideas and talent. Sematech named Robert Noyce as its original president in an effort to supply the administrative savvy needed to ensure cooperation in the highly fragmented microelectronics industry.[66] The inability of U.S. high-technology firms to overcome their differences and develop cooperative strategies ultimately led to the demise of U.S. Memories, a proposed consortia to rebuild U.S. strengths in the field of semiconductor memories. In the words of a leading semiconductor industry executive, the failure of U.S. Memories "is not a very strong endorsement of getting together to solve problems."[67]

R&D consortia are dominated by large established companies, so participation by smaller start-up companies is often difficult. Sematech, for example, counts only two smaller companies, LSI Logic and Micron Technology, among its members. Some entrepreneurs consider Sematech's $2 million entry fee as an attempt to "price out" small companies. Others refuse to participate because they feel that R&D consortia dominated by large firms lag behind the technological cutting edge. Ed Sack, president of Zilog, believes that "Sematech is not particularly relevant to the problems of today. Zilog will not participate."[68]

R&D consortia face technological leaks of the sort associated with the hypermobility of high-tech labor. As Dataquest analyst Sheridan Tatsuno puts it: "Anybody who leaves Sematech would be prime pickings by a Fujitsu or a NEC."[69] Free riders are an obvious problem, and one that is hard to cope with. The permeability of international borders to scientific and technical information make certain that information and know-how developed in U.S. consortia will rapidly become available to overseas competitors. A main conduit for information leakage is the extensive relationships many of the participants have with overseas competitors. Far more through leakage occurs when a foreign competitor purchases a company that is a consortium member, as was the case recently when Japan's Nippon Sanso acquired Semi-Gas Systems, a company that had been involved in and supported by Sematech.

While R&D consortia may produce some benefits, they do not constitute a panacea for American high technology. Although Sematech has generated new technology, it has thus far failed to improve the condition of our semiconductor industry. And it has been forced to bail out a larger and larger number of U.S. companies, a reality reflected in the following *New York Times* headline: "Sematech Today: Cash Dispenser."[70]

The point is simple. R&D consortia are better suited to breakthrough than follow-through. They operate most smoothly when they focus on nonproprietary or "precompetitive" research—R&D that is between basic research and product development. The idea is to develop "generic" sorts of technologies that can be applied in the development and production of a number of different products.[71] There is little reason to expect that American companies will be able to turn the "generic" innovations pioneered by R&D consortia into actual products and processes, especially when they have such a difficult time doing this with their own R&D innovations. In fact, the main reason Japanese consortia are so successful is that corporate participants are geared to rapidly turning R&D into commercial products.[72]

In the end, R&D consortia remain wedded to the "technical fix" mentality of the breakthrough economy. They represent yet another attempt to circumvent the internal organizational changes necessary to rebuild a more integrated system. The search for new breakthroughs will not yield the lessons of follow-through.

The University as "White Knight"

The university has long played a role in capitalism as a trainer of scientists, engineers, and other workers with advanced skills. During the late twentieth century, the university's role has increasingly shifted from teaching to one of basic research.[73] In the years following World War II, the U.S. government became a primary funder of basic research, and it displaced corporations, foundations, and individuals as the largest funders of university research. The primary conduits for this funding were the Department of Defense, the National Science Foundation (NSF), and the National Institutes of Health (NIH). During the postwar era, American universities evolved a basic research establishment that was acknowledged as having no peer. American scientists garnered the lion's share of Nobel prizes. Our universities were meccas to which the world's researchers flocked. Our technological capabilities have been immensely strengthened by the ability of our universities to attract many of the world's best scientists.

With the decline in American competitiveness during the late 1970s and 1980s, concern grew that advances in basic university research were not being exploited by industry. The university came increasingly to be viewed as yet another weapon in the battle for competitiveness. In the 1970s the National Science Foundation established a landmark program to create university-industry centers. And in a major address given in 1980, Derek Bok, president of Harvard University, announced that the university was going to have to ally itself more closely with private industry in an effort to revitalize the American economy.

By the early 1980s a host of new university-industry relations had been forged. These new arrangements are most noticeable in the new biotechnologies where multiyear, multimillion-dollar contracts have been signed between large corporations and major research universities.[74] Washington University in St. Louis is one of a number of universities that has actually allowed venture capitalists to establish an office on campus.[75] And in an even more problematic development, Boston University invested more than $60 million, one-third its total endowment, in a biotechnology start-up, Seragen—an investment blunder that is a major drain on the school's finances.[76] What had previously been unusual liaisons have now become commonplace.

Reconciling the university with the short-term profit motives of industry is difficult at best. While some believe that the university can play an important role in the development of commercial technologies, critics contend that the university is selling out to industry.[77] Many university scientists see the redirection from basic to more applied research as misguided. Robert Park of the American Physical Society states: "We're going to start shoving our universities into more applied and even more developmental work and I think that's a mistake."[78]

The potential for abuse is great and often perpetrated by the universities themselves. When university research comes to be viewed as a source of competitiveness, the norms developed to ensure excellence and reliability of research results are often left behind. By 1989 the rising pressure on universities to deliver had yielded episodes such as the rather unusual involvement of administrators of the University of Utah and officials of the state of Utah in the much-publicized "cold fusion" affair. Here, alleged experimental success in achieving fusion at room temperatures led the state of Utah to appropriate extra funds for the university lab that announced the discovery. These officials descended upon Washington, D.C., demanding special appropriations, using the now-familiar rallying cry of everyone looking for a handout: "The Japanese will surpass us."[79] (The Japanese bogeyman has apparently replaced the Soviet bogeyman of the 1950s, 1960s, and 1970s.) In later months it proved difficult to reproduce the results of the original cold fusion experiment, and the hype eventually receded.

The dilemmas posed by the commercialization of the university are extremely serious. The university is perhaps one of the last major

American institutions that is respected and envied by the rest of the world. Its emphasis on the free exchange and flow of information has created an intellectual commons that has no equal. The university is the institution that produces and reproduces the fundamental scientific and technological skills that are so critical for generating new innovations. The excessive emphasis on commercialization and profitable research has already partially eroded this institutional space. Once the commons is destroyed, it will be impossible to replace.

There is no guarantee that the new university-industry alliances will provide exclusive benefits to American industry. A 1988 study by the U.S. General Accounting Office indicates that 496 foreign corporations now have research relationships with 41 leading U.S. universities. MIT's Industrial Liaison Program has 130 foreign members, including 56 Japanese corporations.[80] Foreign companies are also sending more and more of their scientists on sabbaticals to U.S. universities—a development that has provoked a recent outcry from some American business leaders and policymakers. But such complaints have a hollow ring. The real issue is why these slots are not being taken by American scientists. Either U.S. companies have not been sufficiently aggressive in placing their researchers in university labs, or maybe they do not feel that they would benefit from such endeavors. If they regularly sent their scientists to the university the way they send their managers back to business school, perhaps they might not need to "purchase" university scientists or even entire university laboratories to gain access to university technology.

The real tragedy of the situation is that the root problems of the breakthrough economy have nothing to do with the university. Rather, they lie in U.S. corporations that are poorly organized and therefore cannot turn their innovations into products. A hundred, a thousand, or a million new university innovations will not solve this. It is ill considered to push the university to undertake the economic functions of the industrial corporation. As an institution, the university is simply not equipped to play a significant role in bolstering our weakness in technological follow-through.

Government Steps into the Breach

Government has also gotten involved, with an array of new policies and programs to get the U.S. economy back on track. On the basis of very limited evidence regarding the success of its earlier university-industry centers, the National Science Foundation recently established eighteen new "engineering research centers."[81] As part of the 1988 trade bill, Congress authorized funding for twelve more advanced manufacturing centers to be administered by the National Institute of Standards and Technology (formerly the National Bureau of Standards).[82] Not to be outdone, the National Institutes of Health has now established its own research centers program.

The increase in public funding for R&D consortia is not surprising, in light of the fact that industry-financed R&D has not kept up with the global pace over the past decade. The more R&D is subsidized by the state, the more private capital can be freed up for other activities, such as takeovers, increased dividends, and stock buy-backs. The political imperatives behind the explosion in government-supported technology centers are also clear. These programs offer an opportunity for every university, every senator and representative, to enroll in the crusade to improve U.S. competitiveness. They are a modern twist on the age-old practice of pork barrel funding.

The Defense Department is a key player in the new federal commitment to technology policy. Between 1980 and 1989, defense spending more than doubled, increasing from $140 billion to more than $300 billion.[83] Supporters of an increased Pentagon role see Stars Wars as the centerpiece of America's next thrust in science and technology. On a smaller scale, the Pentagon's Defense University Research Initiative distributes more than $100 million in funds to seventy universities in twenty-nine states.[84] Sematech is funded largely by the military. MCC's initial board chairman was a former CIA operative, Admiral Robert "Bobby" Inman. Many of the members of these two consortia are leading defense contractors. Many are urging the Pentagon to enlarge the scope of its activities to include commercial technology policy. They justify this mainly on the grounds that it would be easy to implement. Some advocates of this strategy

184

are cynical because they know the problems defense funding has caused, but reason that the national security justification is the only way to get funding for the programs they believe are needed.

Supporters of an increased role for the Department of Defense use the misleading term "dual-use" technology to further legitimize this effort. But it is very unlikely that greater defense spending can generate much in the way of civilian spin-offs. For example, consider the following fact. Fewer than one percent of the 8,000 patents from Navy research have been licensed to private corporations, while more than 13 percent of patents generated by the Department of Agriculture have been turned into commercial products. Harvey Brooks places this debate in perspective by observing that "the whole problem of spin-offs from defense is a mess because nobody asks the question: 'Compared to what?' . . . You can't spend $40 billion a year on [defense] R&D without some spin-off. But, it's an accidental by-product. If producing civilian technology is your main objective, and you really don't care much about defense, you could get the same amount of spin-off with much lower expenditures."[85] And dual use is even less likely today than it was before due to recent advances in technology. There are few commercial outlets for specialized defense products such as military semiconductors that must be designed to withstand nuclear battlefield conditions or the technology developed for new Stealth bombers. A 1988 study by the Federal Reserve Bank of Boston indicates that regions receiving large amounts of defense funding do not necessarily generate commercial high technology.[86] Defense and civilian technologies are on separate technological trajectories; hence few commercial spin-offs result from defense R&D.[87] Extending the scope of the Pentagon's authority to include a large measure of control over civilian technology is surely a recipe for disaster.

Worse yet, the increasing reliance on defense technology policy can only impede much-needed corporate restructuring. The major beneficiaries of the Star Wars program are the same old crowd of defense contractors—Lockheed, General Motors, TRW, McDonnell Douglas, Boeing, and General Electric.[88] Between 1980 and 1984, McDonnell Douglas became increasingly dependent on defense outlays as military prime contracts as a percentage of its annual sales skyrocketed from 14 to 67 percent.[89] Extending the scope of the military grants economy only makes it easier for these corporations to ignore the organizational reforms needed to make them truly competitive in commercial markets. And of course, abuse and corruption continue to run rampant in the Pentagon's industrial subsidy system. According to statistics from the Inspector General of the Department of Defense, in late summer of 1988, there were 452 active investigations of fraud, gouging, influence peddling, and other abuses by defense contractors.[90] Computer designer Gordon Bell sarcastically observes that recent military forays into semiconductor technology amount to little more than "public pork barrel" designed to bring a group of "old line miserable semiconductor guys up to the state-of-the-art."[91] He did not add that they probably will not succeed. The rise of military technology policy is tantamount to elevating a major culprit to the position of a savior.

The States Ante Up

State governments have gotten into the act, too, with a wide variety of programs designed to boost the role of state universities in technological innovation and economic growth. Pennsylvania's Ben Franklin Partnership, Ohio's Thomas Edison program, the New York Science and Technology Foundation, the Michigan Strategic Fund, and New Jersey's Advanced Technology Centers are just a few examples of state programs supporting university R&D.[92] And the states have supplemented these efforts with programs for venture capital, entrepreneurial assistance, business incubators, and even science parks.[93] A recent comprehensive study indicates that forty-eight states now sponsor technology programs with expenditures totaling $550 million.[94]

While author David Osborne enshrines these recent state efforts as "laboratories of democracy"—early experiments that will pave the way for later federal and even private sector efforts—there are several good reasons to raise questions about them.[95] Already bidding wars have developed, as states raid each other's universities for directors of their consortia. This has the effect of diluting talent and encouraging a competition for highly desired professors that resembles the competition for sports stars.[96] There are bigger problems of duplication and wasted effort as each state essentially reproduces the policy of the others. These programs run the risk of generating another intense "war between the states" in a mad scramble to attract high-tech companies and jobs.[97]

State venture capital programs may, if anything, be more problematic than those state programs mentioned previously. Unlike private venture capitalists, state venture capital agencies usually do not share in the spoils of big winners but are forced to bear the brunt of losses. What's worse, these programs face a catch-22 situation: states either constrain investment options to local companies, thereby dramatically increasing the possibility of failure, or subsidize private venture capitalists who invest state money in the best deals they can find—usually out-of-state companies. In any event, state venture capital funds go to start-up companies that make a few key individuals very wealthy but create few jobs or benefits for the state economy. And the recent explosion in state high-technology development programs has already spawned a good deal of cronyism, as evidenced in the growing group of "public entrepreneurs" that derives huge profits from these programs. While the states may be able to help at the margin, they are simply ill equipped to solve our underlying technological and industrial problems.

Send in the Lawyers

A recent cartoon in *Datamation* shows an IBM lawyer poring over the rules of baseball and trying to figure out a way for IBM to patent the game. This highlights the dilemmas of yet another highly touted solution to our technological woes: the current wave of intellectual property litigation.[98] Faced with rising competition from Japan and elsewhere, U.S. companies began to seek refuge in the legal system. Under tremendous pressure from high-technology industrialists, Congress passed a number of laws to strengthen intellectual property protection in the 1980s.

The new intellectual property laws are used against domestic as well as foreign competitors, especially against small firms that have either spun off or have hired key personnel from other firms.[99] Cypress Semiconductor, for example, has faced twenty suits or threatened suits since its founder left Advanced Micro Devices.[100] Apple Computer even sued its founder, Steven Jobs, for "misappropriation of trade secrets."[101]

Increasing concern for intellectual property diverts attention from making innovations to protecting them. Many leading high-technology firms have developed large in-house legal staffs for the purpose of protecting their intellectual property. For example, Intel now has eleven highly paid attorneys dedicated solely to intellectual property issues, and the company also hires numerous outside attorneys for litigation. As a result, its total litigation expenses have increased tenfold since 1985, when it had no major intellectual property litigation.[102] DEC has created a huge intellectual property staff of twenty in-house attorneys and has numerous outside law firms on retainer.[103] The situation in biotechnology is even worse: global patent wars are being fought in virtually every product area. Some estimate that high-technology firms now spend in the range of $2 billion per year on trade secret litigation alone.[104] This is a significant diversion of resources away from the critical task of developing new technologies and turning them into products.

Companies frequently use intellectual property rights to retard competitors. Apple Computer, for example, has sued Microsoft Corporation and Hewlett-Packard because their new products use a display that resembles the look of Apple's Macintosh.[105] Lotus and Ashton-Tate are suing smaller rivals for using languages originally developed for their programs.[106] Fox Software, for example, has developed a new program that runs twice as fast as its Ashton-Tate competitor but finds itself charged with infringing on Ashton-Tate's intellectual property. Small start-ups are especially vulnerable to intellectual property litigation since they lack the resources to fight protracted court battles. Lawsuits not only provide the larger company an opportunity to overtake the lead of the small company, but they can also be used to drive away potential investors and customers.[107]

But this strategy has its limits. Although stricter forms of intellectual property protection can bog down start-ups, it is much harder to derail a large Japanese competitor since patents can be invented around, and trade secrets hardly work on products that can be reverse engineered. A survey conducted by economists at Yale University suggests that patents are "highly effective" in only 5 out of 130 industries and "moderately effective" in just 20 others.[108] David Teece observes:

> It has long been known that patents do not work in practice as they do in theory. Rarely, if ever, do patents confer perfect appropriability. . . . Many patents can be "invented around" at modest costs. They are especially ineffective at protecting process innovations. . . . In some industries, particularly where the innovation is embedded in processes, trade secrets are a viable alternative to patents. Trade secret protection is possible, however, only if a firm can put its product before the public and still keep its underlying technology a secret.[109]

Furthermore, the patent game is increasingly Janus-faced. As Japanese companies develop strong patent portfolios of their own, U.S. companies find themselves confronted with formidable adversaries who can retaliate.

A recent patent battle between Motorola and Hitachi provides a striking illustration of this. In January 1989, Motorola sued Hitachi claiming that one of the Japanese company's microprocessors violated a Motorola patent. This action arose out of disputes over a 1986 agreement between the two that allowed each to use some of the other's patents. Hitachi responded aggressively, instructing its staff of patent attorneys to mine its extensive patent portfolio in search of Motorola infringements of Hitachi patents. Hitachi then filed a countersuit within weeks of the original action charging that Motorola had infringed on one of its patents. In October 1989, Hitachi filed three additional intellectual property suits charging Motorola with patent infringement and reverse-engineering its technology. In late March 1990, a U.S. federal court ruled that both companies had committed patent infringements. For its part, Hitachi was ordered to pay Motorola $1.9 million in damages and to stop producing its H-8 microprocessor for the life of Motorola's patent. Hitachi scored a much bigger victory as Motorola was ordered to stop producing its flagship 68030 microprocessor, which is used in the Apple Macintosh series of computers and high-powered workstations made by Sun Microsystems and Hewlett-Packard for the life of Hitachi's patent.[110] Without the 68030, Apple Computer would be forced to shut down much of its production operations. William Lard, associate general counsel for Apple, summarized the impact of the decision on Apple for the court.

> Approximately one-half of Apple's entire computer line uses the Motorola part no. 68030. . . . The 68030 is the heart of Apple's Macintosh IIcx, IIci, IIfx, IIx, and SE/39 computers. There are no substitute microprocessors Apple can use in these products. Without the Motorola 68030, the Apple computers cannot function. . . . Apple has approximately an eight-day supply of 68030s. Without additional 68030s, Apple will not be able to produce half of its products and this will immediately impact on the production employees of Apple. Any disruption in the shipment of Apple computers will undoubtedly have a severe effect on those employed in numerous other industries which furnish products to Apple and which distribute and sell Apple products.[111]

Motorola has estimated that more than 70 companies are dependent on the 68030 and would be forced to shut down production facilities if the microprocessor were no longer available. Fortunately for Motorola and its customers, the company was able to avert disaster by winning a stay of the injunction that allows it to continue producing the microprocessor pending its appeal of its original decision. Still, the lesson for American companies could not be more plain: U.S. firms are finding out that their global competitors are investing sufficient resources in R&D so that they too can effectively play the "patent game."

History highlights the fact that legal walls provide little shelter from powerful overseas competitors. A century and a half ago, in a period with far worse communication and transportation, England failed in desperate attempts to keep its textile innovations out of the hands of "Yankee pirates."[112] Intellectual property protection cannot solve the problems of the breakthrough economy. While legal actions may afford some temporary relief, they cannot compensate for the inability to develop competitive technology and manufacture competitive products. And stricter intellectual property protection may slow down the overall pace of domestic innovation to the point that foreign competitors can catch up. Then the United States will find itself in the unenviable position of being able neither to produce breakthroughs nor to follow through on them.

Spot Repairs Won't Work

While each of these responses suffers from its owns weaknesses, the bigger problem is that they do not fit together as part of a broader solution. These various private programs and public strategies are confused and disorganized, work at cross-purposes, and lack the broad sweep of a sustained collective effort. The litany of responses that American industry and government have come up with so far amounts to little more than a patchwork of solutions that deals with superficial issues but fails to address the deep structural and organizational problems facing U.S. high technology. Such solutions may well afford temporary relief or the appearance of improvement. But absent more fundamental restructuring, the underlying problems of the breakthrough economy will only continue to deepen.

For the better part of a century, the United States has shown an uncanny ability to solve many, if not most, of the technological and economic problems it has faced. But now, for the first time, we are unable to generate the kinds of solutions needed to put us back on track. Current prescriptions treat a variety of symptoms without attending to the underlying disease. It is worse than naïve, it is paralyzing, to think that these spot repairs can save the breakthrough economy. To put it bluntly: patchwork solutions are too little, too late.

10
Transcending the
Breakthrough Economy

■

As we enter the 1990s, a series of changes is drastically reordering the global economy. In this rapidly changing environment, predicating our future on breakthroughs is more precarious than ever before. Simply throwing more money at problems or creating bigger corporations or corporate consortia will not save us if we fail to address the underlying structural, organizational, and institutional problems that lie at the heart of this system of technological development.[1] If we are to transcend the breakthrough economy, it will be necessary to transform many existing institutions and organizations and to create completely new ones.

We will have to uproot many of the ways we structure our society if we are to accomplish such a major shift in our existing economic and social institutions. Such processes of self-renewal occur only seldom in a nation's history. In the United States such an uprooting and redirection last occurred during the New Deal. As we have shown, Japan underwent a similar process, which yielded different results, in the immediate postwar period. Eastern Europe is today experiencing such a transformation, the outcome of which is by no means certain. During these massive upheavals the entire thrust of society and its citizens is aimed at searching out new ways of doing things. Sharp struggles and debates emerge. Older incentives and rules of behavior are transformed and reorganized. And from these there emerge new, stable arrangements and institutions. This process can be costly and the struggles fierce. But it is, nevertheless, superior to a gradual approach. The latter is simply not viable because it remains within the accepted rules of the game, and it is these rules that must be changed. Most ominously, drastic change may become possible only in the face of serious economic collapse.

Our view of the conundrum facing the U.S. differs significantly from the recent spate of reports from elite groups such as the Business Roundtable, the President's Commission on Industrial Competitiveness, the Cuomo Commission on Trade and Competitiveness, and the MIT Commission on Industrial Productivity.[2] Each of these groups offers detailed policies and prescriptions. But all lack a vision of what the future should be, and, as a result, their recommendations resemble the list of patchwork solutions described in chapter 9. Worse yet, most of these proposals ensure that the costs of change will fall most heavily on shop-floor workers, the middle class, and the poor. And none of them recognize the need to mobilize and tap the energy of the people in such a transformation. As such, they will never be able set in motion the economic and social transformation it will take to transcend the breakthrough economy. In fact, they offer only a recipe for further stagnation.

Only a far-reaching institutional transformation can uproot the well-entrenched institutions, methods of organization, rules of behavior, and incentives that make up the breakthrough economy. Despite its serious problems and contradictions, this system remains quite durable because it can confer extraordinary financial benefits on those fortunate ones who participate in it: R&D scientists and engineers, entrepreneurs, and venture capitalists. The breakthrough economy, moreover, is part of a new international order in which the U.S. makes the breakthroughs while others, especially the Japanese, provide the follow-through. In this new global environment, policies or programs that aim to bolster America's breakthrough capacity, such as increasing the flow of U.S. funds to venture capital, will only exacerbate our underlying problems. These efforts are akin to trying to fix a production line that is turning out defective products by speeding up the pace of the line. Nowhere is this new symbiotic relationship more apparent than in the wave of foreign capital flowing directly into U.S. venture capital funds and start-ups. A final, vexing contradiction of the breakthrough economy is now apparent: foreign financiers can effectively perform the role of venture capitalists.

There are three foundations upon which industrial and techno-logical restructuring must be built. First, we must replace the current emphasis on military technology with a comprehensive effort to re-build the civilian technology system and the commercial economy. Second, all workers must be elevated to the position of think-workers and become fully integrated participants in our economy. Third, R&D and manufacturing must be synergistically reintegrated. All of these efforts must, in turn, dovetail with and reinforce a major effort to strengthen social, political, and economic justice and equality. A lack of commitment to such ideals will lead to solutions that are no more than pragmatic Band-Aids that will impede progress and lead to fur-ther stagnation.

Dismantling the Cold War Military Burden

It is absolutely necessary that the United States abandon the military burden of the cold war years. As we have demonstrated, the Penta-gon's version of technology policy has consistently emphasized break-through innovations—great leaps forward with cost as no object. And in so doing, it has contributed massively to the loss of follow-through capacity. Large and small corporations alike have been sheltered by the grants economy—the most expensive industrial subsidy program in all of human history—administered by the Pentagon for more than forty years. Generations of our best engineers and scientists have been taken away from developing technologies with real economic impact. We have wasted enough of our scarce financial and human resources on producing weapons.

A "national security" mentality pervades the entire debate on the future of the U.S. economy. The Pentagon busily issues reports on how it intends to involve itself in "bolstering U.S. industrial competi-tiveness."[3] Even nondefense groups assume that the defense indus-trial base must play a role in both national security and the rebuilding of the commercial economy.[4] Consider the conclusions of the recent report of the MIT Commission on Industrial Productivity:

194

Finally, even if all the economic arguments for the importance of manufacturing were somehow rendered moot, the nation's manufacturing base would still remain fundamentally important to national security. The Department of Defense has estimated that it purchases about 21 percent of the gross product of U.S. manufacturing industries and over a third of the output of high-technology manufacturing industries; it depends on virtually every sector of the manufacturing base for its material. For the nation to become too heavily dependent on foreign technology for its defense would be politically and militarily untenable.[5]

To follow such advice will only accelerate our technological and industrial demise. By now the reasons for this should be obvious. Defense technology has always been problematic in generating commercial spin-offs. It is even more problematic in today's environment of rapid innovation and compressed technology life cycles.[6] What is worse, the military's grants economy operates much like a welfare system for companies that are unable to survive on their own. A constant stream of subsidized defense business enables them to avoid just the kinds of restructuring needed to survive in a globally competitive economy. With the sudden thawing of the cold war, it seems particularly risky and ill timed to use the military as an excuse for industrial policy or as a driver of technological change and economic development. This entire strategy is akin to the belief that the university can save American industry. If our companies want to save themselves, they will have to change. Government or university can only help at the margins.

It will be extraordinarily hard to undo the false connection between defense and technological innovation. Preparations for war have been the driving force behind entire sectors of the U.S. economy and much of postwar policy in science and technology. The "iron triangle" of military contractors, military-related congressional committees, and the Pentagon establishment is a political and economic force of unrivaled power.[7] If this is not enough, the fear of the so-called communist threat is deeply ingrained in huge segments of the American population. Overcoming this complex of forces will be a daunting task.

A continued emphasis on defense and national security can only continue to deplete the U.S. economy and keep American corporations sheltered in the Pentagon's own version of the welfare state. We are the architects of our own demise when we divert investment from pressing technological, industrial, educational, and social upgrading and channel it instead to the military. A strong military will not compensate for a weak economy. The Soviet Union under Gorbachev has awakened to this reality; we alone continue to ignore it.

With the logic of the breakthrough economy in question and the cold war in retreat, the issue of economic reconversion must be returned to the agenda. More than forty years have passed since the truncated debate that followed World War II. This reconversion must be not merely a reconversion from a military to a civilian economy, but it must also address ways of superseding the breakthrough economy. An effective reconversion program must move far beyond the inane debate over the so-called peace dividend, which in its current guise has little to do with either "peace" or a "dividend" in the classic sense of return on productive investment.[8] Can the end of the cold war and the rise of Japan be transformed into an end of the breakthrough economy? What will it take?

It seems to us that a 1989 bill, the Defense Adjustment Act, authored by Congressman Ted Weiss of New York, would begin the process of change.[9] This bill has indeed generated considerable support in the Congress; it has already been cosponsored by more than fifty members and has earned the number H.R. 101 to indicate the seriousness with which it is taken in the 101st Congress of the United States.[10] The bill mandates that cities and states should use the freed-up funds to reinvest in our decayed infrastructure and to retrain and reemploy workers. It would create a new Defense Economic Adjustment Council cochaired by the secretaries of commerce and labor to carry out this reconversion effort. But even such a comprehensive effort to rebuild our industrial plant and civilian public works will fail if it simply repeats past mistakes, that is, if it is not part of a broader strategy to remedy the deep structural and organizational problems that face us. Reconversion cannot be undertaken under the auspices of the Pentagon or with significant participation by defense contractors with their existing managements: they have already failed in attempts to reconvert defense installations under the most propitious conditions.*

An effective reconversion program must be combined with new, more decentralized methods of administration and governance. The adage to think globally, act locally must give way to the new one—act locally, connect globally. Local groups of citizens and workers must be formed into regional, even national, networks to provide input into firm-level, industry-level, and government-level decision making. This will ensure that decisions are made in the interest of and on the basis of knowledge provided by those who have the most to lose from them. It will also provide an important check against corporate and governmental irresponsibility of the sort displayed in the recent junk bond fiasco or the savings and loan debacle and subsequent bailout.

There are examples of what can be done. For example, workers and engineers at the British firm Lucas Aerospace developed a program to diversify their company from military to consumer products that were socially beneficial. This released the creativity of both workers and researchers, and new possibilities for economic growth were developed. However, this promising new direction was anathema to the old-fashioned English management and was rejected out of hand.[11] In the current environment, it is very likely that America's leaders would react similarly.

Reconversion essentially means a fundamental reexamination of the logic and basis of our major institutions, our economy, and our society. Can the old forms of organization and shibboleths of the past provide the concepts and meanings for the future? Can a huge part of our population continue to occupy itself preparing for war? The answer is no. An effective reconversion program must provide both the logic and the basis for transcending the breakthrough economy.

---

*A study of six so-called conversion projects undertaken by the Defense Department's Office of Economic Adjustment found that all the commercial products developed were poorly built and too expensive and that all failed. This Pentagon office trumpets the transformation of an abandoned air force base in Michigan into a prison as among its most successful efforts. See John Ullmann, "Economic Conversion: Indispensable for America's Economic Recovery" (Briefing Paper no. 3, National Commission for Economic Conversion and Disarmament, Washington, D.C., April 1989).

The Shop Floor: A Critical Arena of Innovation

Workers must achieve a new role in both the economy and society. Numerous authors have commented on the necessity of having "smart workers" on "smart machines."[12] What they more often neglect is that committed workers are as important as smart workers. The most competitive firms and the most competitive nations invariably provide the kinds of employment guarantees it takes to ensure that workers are committed workers. Yet U.S. corporations and American society continue to see workers as expendable to protect short-term profits. This is true in both the industrial behemoths of the old follow-through economy and the high-tech start-ups of the new breakthrough economy, where the shop-floor worker is considered either a recalcitrant obstacle or a necessary nuisance. A smart worker who understands that he or she is expendable can never be a committed worker. American workers are resisting and have every right to resist corporate efforts to tap into their "brains" if they are not given real guarantees that their jobs are safe.

The sad history of shop-floor workers in advanced industrial countries has been that if they do not organize and demand a better deal they will not receive it. As has recently been demonstrated in Eastern Europe, the energy and excitement of average citizens can be the catalyst for massive restructuring. We require such an upsurge to put new shop-floor power relations into place and in the process break down the old worn-out institutions of the past. At a minimum, U.S. workers should demand joint consultation such as occurs in Sweden or West Germany and the ability to control their immediate work environment, beyond the limited employee input found in Japan. A system of long-term employment is also important, as it would change the cost-benefit calculus of corporations regarding activities such as training and it would enhance the importance of worker initiative on the shop floor. Perhaps it would also help to reorient industry away from its current short-term approach. Further, a long-term commitment to workers and an integrated role for workers in all of the activities of the firm might prove to put a damper on takeover activities—scaring off raiders who seek to profit by closing plants and firing workers. Employees must demand that the firm make the employment contract a serious commitment, not a decision that is easily reversed.

In essence, employees will have to force top management to allow them to have input into the firm. The input must be accompanied by both responsibilities *and* rights. Employees must demand that the firm operate in such a manner as to discharge its responsibility to those who give their efforts to make it a success.

Reconnecting the Factory and the R&D Laboratory

Tapping the intelligence of the entire spectrum of employees from R&D to manufacturing is essential for success in an age of perpetual innovation.[13] Workers must be able to use their intelligence to fashion product and process improvements. The U.S. cannot create a globally competitive high-technology system until the artificial barriers between R&D laboratories and shop floors are abolished. Contrary to the current situation of one-way information flow from the "all-knowing" research scientists and engineers to the "know-nothings" on the shop floor, a true interchange of information is necessary.

At the macro level, corporations need to eliminate both the physical and the social walls separating R&D and manufacturing. They need to overcome the extreme spatial division of labor between the sites of innovation and production and begin to link these two critical activities more closely together. They must also overcome internal divisions between R&D and manufacturing in ways that create synergies between them.

At the micro level, engineers and shop-floor workers must be encouraged to interact and cooperate on a day-to-day level. Here, a useful example of what we need to do can be found in our own industrial past. During the formative years of Silicon Valley, companies such as Hewlett-Packard encouraged close interaction between engineers and shop-floor workers. Engineers would work closely with machinists and shop-floor workers to come up with better designs and better products. Art Fong, one of the first Hewlett-Packard engineers, explains: "In most companies, engineers draw up a schematic and hand it over to a technician who would try to implement it. HP would never stand for that. At HP, we [the engineers] had no special status. We got on the machines ourselves and tried out our ideas." Fong continues: "We knew the men on the production line on a first-name basis. One of them would come into the lab and say, 'Art, this thing is messed up.' And you listened. A good machinist recognizes problems. He's a damn good craftsman with damn good ideas on how to make things simpler and more reliable."[14] A few companies like Hewlett-Packard have managed to preserve this legacy, but most either never had it or forgot it.

To link the factory and the R&D laboratory effectively will require a new worldview that sees all workers as think-workers. The rigid demarcations between white- and blue-collar workers must be broken down and this will probably only happen because blue-collar employees insist on being included in the firm. In fact, their demands for inclusion and a long-term commitment may be echoed increasingly by white-collar managers and engineers who are also finding themselves on the block as U.S. firms undertake cost-cutting measures. In effect, the U.S. must "white-collarize" all of its workers. Shop-floor workers must become equals within the company. The shop floor can then also become a laboratory; and the laboratory must be seen as an extension of the shop floor. Rather than stifling creativity, it is necessary to nurture it wherever it is found. The lab and the shop must be linked in the seamless web of production.

An industrial system that deliberately creates impenetrable boundaries between different aspects of the overall production process will naturally be a victim of its separations. In an era of continuous innovation, hybrid "fusion" technologies, decreasing time to markets, and knowledge-intensive industries, unnecessary divisions will produce a competitive disadvantage. Unification will require empowering those without power. Shop-floor workers must be able to raise suggestions to management and R&D without fear that they will be fired. There must be a constant dialogue between R&D scientists, management, and factory workers. But, again, only through empowerment can a productive synthesis emerge.

Building Networks of Firms

New types of relationships between firms can also play a role in reversing our technological and industrial decline. Long-term relationships of mutual economic benefit are necessary to harness our technological and industrial strength, especially when so many of our high-technology companies are small. But these will be very difficult to establish in the hypercompetitive U.S. environment, where the turnover of people and even firms is extraordinarily high and where cutthroat competition and short-term goals predominate. American high-technology corporations continue to operate under the assumption that relationships between firms are essentially competitive relationships between independent parties—each aiming to maximize its own profits with little regard for others. In effect, U.S. industry does not recognize the fact that the entire production chain is interconnected and that, therefore, failure at any link will lead to failure of the entire chain.

As we have seen, American corporations have a long legacy of squeezing their suppliers even to the point of bankruptcy to supplement and bolster their own profits. Suppliers frequently respond to this situation by using substandard materials, cutting back on workmanship, and doing everything they can to cut costs. This can and does lead to a vicious cycle of worsening relationships, lower-quality products, deteriorating manufacturing capabilities, and ultimately a shrinking base of suppliers. In Silicon Valley and to a lesser extent in the Route 128 area around Boston, this has created a situation in which American high-technology corporations are increasingly dependent upon foreign suppliers, who are simultaneously important competitors, for critical inputs into the production process.

Although there is no substitute for long-term relationships, these cannot be created in an environment of extreme turmoil or where short-term goals predominate. Corporations that are unable to make long-term commitments to their own employees will find it extremely difficult to make such commitments to other firms. At the most basic level, relationships between firms (and even different parts of one firm) are always based on individuals. The constant turnover of personnel—and increasingly the buying and selling of divisions of firms and entire firms—constantly disrupts these personal linkages, thwarting any long-term benefits that might accrue from them. In this high-turnover, hypermobile environment it is virtually impossible to establish integrated "obligational" networks. Without the fundamental changes recommended earlier, truly synergistic networks of firms—large and small alike—will remain an impossibility.

Advance or Retreat

The breakthrough economy cannot ensure the economic health of our society. The organizational structures and economic strategies that have brought us to this impasse are no longer able to move our economy forward. New visions, new passions, are needed to re-awaken us from the current malaise, transform outmoded institutions, and reestablish a sense of purpose in our economy and society. While our social, political, and economic system deteriorates, corporate leaders continue to manage their companies on the basis of portfolio investment aimed at the short run, and politicians concern themselves with flag burning, petty corruption, and denying women the freedom of choice. Our citizens yawn as they are bombarded with account after account of the decline of American industry and even their own falling standard of living.

Even the most well-meaning commissions made up of corporate executives, elected officials, and labor union leaders cannot galvanize the broad grass roots effort that is required. And such an effort will never be mobilized by cynical politicians who support democracy in Eastern Europe while assisting in the repression of workers in the U.S. and by business leaders who ask for a worker's involvement in the firm while preparing for mass layoffs. When vision is mistaken for turning a profit on the next quarterly report or a blip in the opinion polls, little can be expected from common citizens.

The energy of workers and citizens is needed to transform our system and build new institutional structures. The past provides certain grounds for optimism. Our nation has so much diversity, so much potential that can be released to overturn the present apathy and cynicism and once again fashion new development trajectories. Employees, consumers, citizens concerned about our environment, and the poor continue to register their dissatisfaction with the current state of affairs and demand full participation in this society. However, these groups have not yet come together in a way that forces others to listen to them and take them seriously instead of simply dismissing or resisting them.

We can move forward or we can stagnate. To follow our current path will mean a slow, steady decline punctuated by more radical drops associated with the business cycle. As this occurs, the U.S. will become a second-rate economy that cannot deliver economic opportunities for the vast majority of its people or the social welfare of Western European countries. We can then expect an ever-increasing malaise and depression in our R&D laboratories, ennui and apathy in our factories, and increased crime in our streets. But if we can break with the assumptions of both the old follow-through economy and the breakthrough economy, we may achieve a new synthesis that can develop new technologies and harness them in ways that will improve the living standards, not only of the privileged, but especially of those who have been denied the benefits by our system.

Notes

■

Chapter 1

1. See U.S. Congress, Office of Technology Assessment, *Commercializing High-Temperature Superconductivity* (Washington, D.C.: U.S. Government Printing Office, 1988); "U.S. Superconductivity Effort Shapes Up," *Physics Today* (August 1989):55–57; Simon Foner and Terry Orlando, "Superconductors: The Long Road Ahead," *Technology Review* (February–March 1988):46–47; and David Gumpert, "The Big Chill," *Electronic Business* (January 23, 1989):24–32.
2. See, for example, the journal *Cambridge Report on Superconductivity* (Cambridge, Mass.), various issues, for information on the various start-up activities in this field.
3. Our discussion of American Superconductor Corporation and the superconductor industry in general is based upon extensive interviews with Yurek, Vander Sande, McKinney, and other principals and financial backers of American Superconductor Corporation. David Talento, a former graduate student at Carnegie Mellon University, conducted these interviews in April 1989. See Richard Florida and David Talento, "Venture Capital's Role in Commercializing University Technology: The Case of American Superconductor Corporation" (Discussion Paper, School of Urban and Public Affairs, Carnegie Mellon University, Pittsburgh, Pa., January 1990).
4. See Robert Pool, "Keeping Up with the Jonezawas," *Science* 245 (August 1989):594–95.
5. Even though U.S. companies were invited to join, only two accepted limited affiliate memberships. See "Japan Keeps a Lock on Its Superconductor Labs," *Business Week,* September 19, 1989, 76–78; and "U.S. Technology Experts Assess Japan's Superconductivity Effort," *IEEE News Supplement* (October 1989).

6. See U.S. Congress, Office of Technology Assessment, *Commercializing High-Temperature Superconductivity;* and Andrew Pollack, "U.S. Reported Trailing Japan in the Superconductor Race," *New York Times,* October 16, 1988, 1, 9.

7. Principals of American Superconductor, interview by David Talento, April 1989, including Gregory Yurek and John Vander Sande of American Superconductor, George McKinney of American Research and Development, and Lita Nelson of the MIT Licensing Office.

8. The story of the U.S. biotechnology effort is chronicled in Martin Kenney, *Biotechnology: The University-Industrial Complex* (New Haven: Yale University Press, 1986).

9. On the four tigers, see "The Four Tigers Are Maturing Quickly," *Electronic Business* (December 11, 1989):89–90. A comprehensive account of South Korea is provided in Alice Amsden, *Asia's Next Giant: South Korea and Late Industrialization* (New York: Oxford University Press, 1989).

10. On the U.S. decline in semiconductors, see Thomas Howell, William Noellert, Janet MacLaughlin, and Alan Wolff, *The Microelectronics Race: The Impact of Government Policy on International Competition* (Boulder, Colo.: Westview Press, 1988); Michael Borrus, *Competing for Control: America's Stake in Microelectronics* (Cambridge, Mass.: Ballinger, 1988); and U.S. Defense Science Board, *Report of the Defense Science Board Taskforce on Defense Semiconductor Dependency* (Washington, D.C.: U.S. Department of Defense, 1987).

11. On computers, see George Gamota and Wendy Frieman, *Gaining Ground: Japan's Strides in Science and Technology* (Cambridge, Mass.: Ballinger, 1988), 23–40; and Willie Schatz, "IEEE Warns of the Japanese Supercomputer Threat," *Datamation* (August 15, 1988); "After a Shakeout U.S. Disk Makers Face Japan," *New York Times,* January 21, 1990, Section 3, 10; and "The Japanese Threat in Mainframes Has Finally Arrived," *Business Week,* April 9, 1990, 24.

12. See "Japanese Drive Makers Set Sights on 3.5-Inch Market," *Electronic Business* (August 21, 1989):17–18; and Joel Dreyfuss, "Getting High Tech Back on Track," *Fortune,* January 1, 1990, 74–77.

13. See Gamota and Frieman, *Gaining Ground,* 107–30; and Japan Economic Institute, "Japan's Telecommunications Market: 1989 Update," *JEI Report,* no. 46A (December 8, 1989).

14. See Michael Cusumano, "The Factory Approach to Large-Scale Software Development: Implications for Strategy, Technology and Structure" (Working Paper no. 1885–87, MIT Sloan School of Management, Cambridge, 1987). In fact, Ronald Rohrer, a leading software expert, believes that Japan's ability to manufacture reliable software may allow it to overtake a huge U.S. lead by the early twenty-first century. Personal communication with Richard Florida.

15. See Gamota and Frieman, *Gaining Ground*, 131–54; and Mark Dibner and R. White, "Biotechnology in the United States and Japan," *Biopharm* 2, 2 (1989):22–29.

16. T. Kitayama quoted in "In Computer Research Race Japanese Increase Their Lead," *New York Times*, February 21, 1990, C4; and Michael Liebowitz, "X-ray Lithography: Wave of the Future," *Electronic Business* (November 27, 1989):28–35; also Gamota and Frieman, *Gaining Ground*.

17. Receipts from royalties and license fees sold to Japanese corporations were $1.3 billion in 1986, $1.9 billion in 1987, and $2.4 billion in 1988. U.S. Department of Commerce, Bureau of Economic Analysis (Unpublished data, January 1990).

18. This is the cumulative total of all royalties and fees paid to U.S. corporations between 1980 and 1989. Data are from U.S. Department of Commerce, Bureau of Economic Analysis, *Survey of Current Business* (Washington, D.C.: U.S. Government Printing Office, June 1989) for 1980–88, and (September 1989) for data on the first three-quarters of 1989, which have been estimated for the entire year.

19. Data on company patents is from Francis Nairn and Dominic Olivastro, "Identifying Areas of Leading Edge Japanese Science and Technology," *Report to the National Science Foundation* (Washington, D.C.: National Science Foundation, 1988).

20. These data are for fiscal year 1990 as compiled by U.S. Department of Commerce, Patent and Trademark Office, "Technology Assessment and Forecast Database" (Unpublished data, January 1990).

21. Database technology was the one area where they thought Japan was clearly behind. Results of a Ministry of International Trade and Industry (MITI) survey of Japanese executives as reported in Japan Economic Institute, "MITI: Industrial Technology Tops," *JEI Report*, no. 37B (September 30, 1988). The MITI report is also discussed in Paula Doe, "MITI Report Pats Japanese Technologists on the Back," *Electronic Business* (May 29, 1989):63–64.

22. Results of a poll of top executives and R&D scientists at the largest 476 U.S. companies. See "Nation's CEOs Say U.S. May Lose Its Research and Development Lead," *Research and Development Magazine* (September 1988):17–18. Results of a poll of 750 CEOs as reported in Bruce Rayner, "How U.S. Electronics CEOs Will Address the New Competitive Priorities," *Electronic Business* (March 19, 1990):34–39.

23. Peter Brooke, interview by authors, June 1987.

24. Henry Ergas of the Organization for Economic Cooperation and Development (OECD) provides a useful comparative framework from which to compare the U.S., Japan, and other national systems of technology development. According to his framework, the U.S. is extremely good at moving into new technological frontiers, or "shifting." He contrasts

this with West Germany, which gets the most out of old technologies, what he calls "deepening." Japan, however, has developed a new kind of technology system that simultaneously "shifts" and "deepens." See Henry Ergas, "Does Technology Policy Matter?" in Harvey Brooks and Bruce Guile, eds., *Technology and Global Industry* (Washington, D.C.: National Academy Press, 1987), 191–245.

25. Clyde Prestowitz, *Trading Places: How We Allowed Japan to Take the Lead* (New York: Basic Books, 1988). The classic citation on Japanese industrial policy is Chalmers Johnson, *MITI and the Japanese Miracle* (Stanford, Calif.: Stanford University Press, 1982).

26. See Robert Reich, *Tales of a New America* (New York: Times Books, 1987); and Charles Ferguson, "From the People Who Brought You Voodoo Economics," *Harvard Business Review* 66 (May–June 1988):55–62. See also "High Tech Entrepreneurs: New Doubt on a U.S. Ideal," *New York Times,* June 14, 1988, 1.

27. A defense of size as an organizational characteristic is advanced in Bennett Harrison, "The Big Firms Are Coming out of the Corner: The Resurgence of Economic Scale and Industrial Power in the Age of Flexibility" (SUPA Discussion Paper, Carnegie Mellon University, Pittsburgh, Pa., September 1989).

28. See Rick Whiting, "Whither IBM: Can the Giant Reassert Itself?" *Electronic Business* (December 11, 1989):112–14; Marcia Berss, "Bull in a Bear Cage," *Forbes,* March 5, 1990, 78–80.

29. On the problems of U.S. manufacturing, see Stephen Cohen and John Zysman, *Manufacturing Matters: The Myth of the Post-Industrial Economy* (New York: Basic Books, 1987). On our sagging product development capability, see Robert Hayes, Steven Wheelwright, and Kim Clark, *Dynamic Manufacturing: Creating the Learning Organization* (New York: Free Press, 1988). An informative discussion on the economic gains that can be obtained by integrating various organizational functions (e.g., R&D and marketing) is provided in Edwin Mansfield, *Industrial Research and Technological Innovation* (New York: Norton Press, 1968).

30. On the distinctions between product and process innovation, see William Abernathy and James Utterback, "Patterns of Industrial Innovation," *Technology Review* (June–July 1978):41–47. See also Frank Hull, Jerald Hage, and Koya Azumi, "R&D Management Strategies: America versus Japan," *IEEE Transactions on Engineering Management* 32, 2 (May 1985):78–83, for an illuminating discussion of the relationship between organizational characteristics and type of innovation.

31. On these types of innovations, see Fumio Kodama, "Japanese Innovation in Mechatronics Technology," *Science and Public Policy* 13, 1 (February 1988).

32. On systems innovations, see Thomas Hughes, *Networks of Power* (Baltimore, Md.: Johns Hopkins University Press, 1983), chaps. 1, 2.

Chapter 2

1. George Perazich and Philip Field, *Industrial Research and Changing Technology* (Philadelphia, Pa.: Works Progress Administration, 1940), 42–43.

2. In a recent paper, William Lazonick of Columbia University provides insight into the relationships among the various elements of the modern corporation. See William Lazonick, "Business Organization and Competitive Advantage: Capitalist Transformations in the Twentieth Century" (Unpublished paper, Department of Economics, Columbia University, New York, September 1988). For an interesting discussion of U.S. decline, see Richard Nelson, "U.S. Technological Leadership: Where Did It Come from and Where Did It Go?" *Research Policy* 19 (1990): 117–132.

3. On Taylorism and scientific management, see Daniel Nelson, *Managers and Workers* (Madison: University of Wisconsin Press, 1975); idem, *Frederick W. Taylor and the Rise of Scientific Management* (Madison: University of Wisconsin Press, 1980). On Fordism, see Michel Aglietta, *A Theory of Capitalist Regulation: The U.S. Experience* (London: New Left Books, 1979); and David Hounshell, *From the American System to Mass Production* (Baltimore, Md.: Johns Hopkins University Press, 1984).

4. See especially Harry Braverman, *Labor and Monopoly Capitalism* (New York: Monthly Review Press, 1974); David Montgomery, *Workers' Control in America* (New York: Cambridge University Press, 1979); and Richard Edwards, *Contested Terrain* (New York: Basic Books, 1979). There has recently been a major reevaluation of the Braverman hypothesis; see, for example, Shoshana Zuboff, *In the Age of the Smart Machine* (New York: Basic Books, 1989).

5. See the classic works of Alfred Chandler: *Strategy and Structure: Chapters in the History of Industrial Enterprise* (Cambridge: MIT Press, 1962); *The Visible Hand* (Cambridge: Harvard University Press, 1977); and *Scale and Scope: The Dynamics of Industrial Capitalism* (Cambridge: Harvard University Press, 1990).

6. Recent years have seen an outpouring of excellent books and articles on the historical evolution of industrial research. Especially useful are the following: David Noble, *America by Design: Science, Technology and the Rise of Corporate Capitalism* (New York: Oxford University Press, 1977); David Hounshell, ed., *The R&D Pioneers* (forthcoming); Margaret Graham, "Industrial Research in the Era of Big Science," *Research in Technological Innovation, Management and Policy* 2 (1985):47–79; David Mowery and Nathan Rosenberg, *Technology and the Pursuit of Economic Growth* (Cambridge: Cambridge University Press, 1989). A comprehensive overview of recent research is provided in Michael Dennis, "Accounting

for Research: New Histories of Corporate Laboratories and the Social History of American Science," *Social Studies of Science* 17 (1987):479–518.

7. Moreover, Schumpeter found this new corporate form so powerful that he completely revised his earlier theory, which had emphasized the importance of "exceptional entrepreneurs" in technological advance. See Joseph Schumpeter, *Capitalism, Socialism and Democracy* (New York: Harper, 1975). Consider John Kenneth Galbraith's opinion: "There is no more pleasant fiction than that technical change is the product of the matchless ingenuity of the small man forced by competition to employ his wits to better his neighbor. Unhappily it is a fiction. Because development is costly, it follows that it can be carried out only by a firm that has the resources associated with considerable size." *American Capitalism* (Boston: Houghton Mifflin, 1956), 86.

8. On the evolution of Edison's efforts from Menlo Park to General Electric, see George Wise, "R&D at General Electric, 1878–1985," in Hounshell, ed., *R&D Pioneers.*

9. Cited in Thomas P. Hughes, *American Genesis: A Century of Invention and Technological Enthusiasm* (New York: Viking, 1989), 29.

10. Indeed, Norbert Weiner described the significance of Edison's laboratory this way: "Edison's greatest invention was that of the industrial research laboratory. . . . The GE Company, the Westinghouse interests and the Bell Telephone Labs followed in his footsteps, employing scientists by hundreds where Edison employed them by tens." Quoted in Noble, *America by Design,* 113.

11. Data on the growth of corporate R&D laboratories are from David Mowery, "The Emergence and Growth of Industrial Research in American Manufacturing, 1899–1945" (Ph.D. dissertation, Stanford University, 1981). Detailed historical discussions of early corporate laboratories are provided in the following. For a comprehensive collection of articles on this issue, see David Hounshell, ed., *The R&D Pioneers.* See also Leonard Reich, *The Making of Industrial Research: Science and Business at GE and Bell, 1876–1926* (New York: Cambridge University Press, 1985); Stuart Leslie, *Boss Kettering: Wizard of General Motors* (New York: Columbia University Press, 1983); and Margaret Graham, *RCA and the VideoDisc: The Business of Research* (New York: Cambridge University Press, 1986).

12. Quoted in George Wise, *Willis R. Whitney, General Electric, and the Origins of U.S. Industrial Research* (New York: Columbia, 1985), 77.

13. Quoted in David Hounshell and John Smith, *Science and Corporate Strategy: DuPont R&D, 1902–1980* (New York: Cambridge University Press, 1988), 45.

14. For a theoretical perspective, see Wesley Cohen and Daniel Levinthal,

"Innovation and Learning: The Two Faces of R&D," *Economic Journal* 99 (September 1989):569–96.

15. Quoted in Thomas P. Hughes, *Networks of Power: Electrification in Western Society, 1880–1930* (Baltimore, Md.: Johns Hopkins University Press, 1983), 105.

16. See Leonard Reich, "Research, Patents and the Struggle to Control Radio."

17. See David Mowery and Nathan Rosenberg, *Technology and the Pursuit of Economic Growth* (Cambridge: Cambridge University Press, 1989); George Perazich and Philip Field, *Industrial Research and Changing Technology* (Philadelphia, Pa.: Works Progress Administration, 1940). It should be noted that by the time of World War I, Germany displaced Britain, to become the second most powerful industrial economy. The Germans excelled in the chemical and electrical industries, an advantage that they have retained until the present.

18. For example, J. E. Walters, *Research Management: Principles and Practice* (Washington, D.C.: Spartan Books, 1965), 110–12.

19. See Graham, *RCA and the VideoDisc,* 72.

20. Interviews by Richard Florida at Westinghouse corporate research center, June–July 1988.

21. Noble, *America by Design,* 118.

22. J. A. Morton, a former Bell Labs researcher, as quoted in *Organizing for Innovation* (1971):63.

23. George Wise, "Science and Technology," *OSIRIS* 2, 1 (1985):229.

24. Schumpeter, *Capitalism, Socialism and Democracy,* 132–33.

25. Donald Miller, *Managing Professionals in R&D* (San Francisco: Jossey-Bass, 1986), ix.

26. This is part of what Rosabeth Kanter calls "segmentalism." See *The Change Masters: Innovation and Entrepreneurship in the American Corporation* (New York: Simon & Schuster, 1983).

27. Jack Goldman, "Can Industry Afford Basic Research," in National Conference on Industrial Research, *New Products and Profits: Proceedings of the Fourth Annual National Conference on Industrial Research* (Beverly Shores, Ind.: Industrial Research Inc., 1969), 38.

28. Emmanuel R. Piore, "The Function of Research in a Corporation or Industry," in Dean Morse and Aaron Warner, eds., *Technological Innovation and Society* (New York: Columbia University Press, 1966).

29. Robert Hayes and William Abernathy, "Managing Our Way to Economic Decline," *Harvard Business Review* 58 (July–August 1980):67–77.

30. The impact of the military on the management of industrial research is discussed in Merritt Roe Smith, ed., *Military Enterprise and Technological Change* (Cambridge: MIT Press, 1985), especially the article by David Allison, "U.S. Navy Research and Development since World War II."

See also Paul Forman, "Behind Quantum Electronics: National Security as a Basis for Physical Research in the United States, 1940–1960," *Historical Studies in Physical and Biological Sciences* 18, 1 (1987):149–229.

31. As quoted in H. W. Coover, "Programmed Innovation: Strategy for Success," in Ralph Landau and Nathan Rosenberg, eds., *The Positive Sum Strategy: Harnessing Technology for Economic Growth* (Washington, D.C.: National Academy Press, 1986), 402.

32. On this new locational strategy, see Stephen Hymer, "The Multinational Corporation and the Law of Uneven Development," in J. Bhagwati, ed., *Economics and World Order* (New York: Free Press, 1972), 141–58; and Robert Cohen, "The New International Division of Labor, Multinational Corporations and Urban Hierarchy," in Michael Dear and Allen Scott, eds., *Urbanization and Urban Planning in Capitalist Society* (New York: Methuen, 1981), 287–318.

33. See Doreen Massey, *Spatial Divisions of Labor: Social Structure and the Geography of Production* (New York: Methuen, 1984). For a critique of this approach, see Michael Storper and Richard Walker, *The Capitalist Imperative: Territory, Technology, and Industrial Growth* (London: Basil Blackwell, 1989). For an empirical analysis of the location of R&D facilities, see Edward Malecki, "Research and Development and the Geography of High-Tech Complexes," in J. Rees, ed., *Technology, Regions and Policy* (Totawa, N.J.: Rowman & Littlefield, 1986), 51–73.

34. This example is drawn from personal interviews with Westinghouse R&D scientists and executives conducted by Richard Florida at Westinghouse corporate R&D labs, June–July 1988.

35. See Graham, *RCA and the VideoDisc,* 71.

36. Graham, "Industrial Research in the Age of Big Science," 75–76.

37. See Graham, *RCA and The VideoDisc.*

38. See Hounshell and Smith, *Science and Corporate Strategy.*

39. See, for example, Edward Roberts, "New Ventures for Corporate Growth," *Harvard Business Review* 58 (July–August 1980):134–42; and Norman Fast, "Pitfalls of Corporate Venturing," *Research Management* 24, 2 (1981):21–24.

40. See especially Douglas Smith and Robert Alexander, *Fumbling the Future: How Xerox Invented, Then Ignored the First Personal Computer* (New York: William Morrow, 1988); George Pake, "Research at Xerox PARC: A Founder's Assessment," *IEEE Spectrum* (October 1985):54–61; and Tekla Perry and Paul Wallich, "Inside the PARC: The 'Information Architects'" *IEEE Spectrum* (October 1985):62–75.

41. Gary Jacobson and John Hillkirk, *Xerox: American Samurai* (New York: Collier Books, 1986).

42. Quoted in Perry and Wallich, "Inside the PARC," 73.

43. On the Alto case, see Robert Burgelman and Leonard Sayles, *Inside Corporate Innovation* (New York: Macmillan, 1986), 19.

44. Quoted in John Sculley, *Odyssey: Pepsi to Apple* (New York: Harper & Row, 1987), 206.

45. The rise of the defense economy has been chronicled in a wide variety of books and articles. But the earliest and most perceptive critic has been Seymour Melman. The all-pervasive nature of defense spending since World War II fully justifies the observation that national defense has been the U.S. industrial policy in the postwar period. See Seymour Melman, *The Permanent War Economy* (New York: Simon & Schuster, 1974); and idem, *Pentagon Capitalism: The Political Economy of War* (New York: McGraw-Hill, 1970).

46. See Graham, "Industrial Research in the Age of Big Science," 53.

47. Melman, *Pentagon Capitalism*, 231–34.

48. See Melman, *Permanent War Economy*.

49. See Nathan Rosenberg, "Civilian 'Spillovers' from Military R&D Spending: The U.S. Experience since World War II," in Sanford Lakoff and Randy Willoughby, eds., *Strategic Defense and the Western Alliance* (Lexington, Mass.: Lexington Books, 1987), 165–88.

50. This figure is from Seymour Melman, "What to Do with the Cold War Money," *New York Times,* December 17, 1989, F3.

51. See George Wise, "R&D at General Electric, 1878–1985" (Prepared for the R&D Pioneers Conference, Hagley Museum and Library, Wilmington, Delaware, October 7, 1985), 24.

52. Reuben Mettler, interview in *High Technology Business* (September 1987):52.

53. See David Noble, *Forces of Production: A Social History of Industrial Automation* (New York: Knopf, 1984); and Max Holland, *When the Machine Stopped* (Boston: Harvard University Business School Press, 1989).

54. The historian Thomas Misa concludes that dependence upon defense funding caused many of our electronics giants to fall far behind entrepreneurial companies (who received far less in defense subsidies for research), leading him to conclude that "spillover from military to commercial uses was incomplete at best." See Thomas Misa, "Military Needs, Commercial Realities and the Development of the Transistor, 1948–1958," in Merritt Roe Smith, ed., *Military Enterprise and Technological Change*, 253–88.

55. See Rosenberg, "Civilian 'Spillovers' from Military R&D Spending," for a discussion of the limited dual-use impacts of military technology. See also "Re-thinking the Military's Role in the Economy: An Interview with Harvey Brooks and Lewis Branscomb," *Technology Review* (August–September 1989):55–64.

56. James Utterback and Albert Murray, *The Influence of Defense Procurement and Sponsorship of Research and Development on the Development of the Civilian Electronic Industry* (Cambridge: Center for Policy Alternatives, MIT, 1977), 2.

57. As quoted in Robert Krinsky, "Swords into Sheepskins: Militarization of Higher Education in the United States and Prospects for Its Conversion," in Lloyd Dumas and Mazek Thee, eds., *Making Peace Possible: The Promise of Economic Conversion* (Elmsford, N.Y.: Pergamon Press, 1989), 96.

58. See Frank Lichtenberg, "Crowding Out: The Impact of the Strategic Defense Initiative on U.S. Civilian R&D Investment and Industrial Competitiveness" (New York: Columbia University Graduate School of Business, 1988). According to Seymour Melman, crowding out entails "the cost to the community of the opportunity for, and the results from, the productivity improvement that is necessarily forgone by using up resources for military products that might otherwise be used to fabricate and operate new means of production." Melman, *Profits without Production* (Philadelphia: University of Pennsylvania Press, 1983), 85.

59. See James Conant, *Modern Science and Modern Man* (New York: Anchor, 1953), 53.

Chapter 3

1. Quoted in T. R. Reid, *The Chip* (New York: Simon & Schuster, 1984), 191.

2. Tandem Computers, "Hiring at Tandem," *Center* (Winter 1986):21.

3. On the early evolution of the computer industry, see Kenneth Flamm, *Creating the Computer: Government, Industry and High Technology* (Washington, D.C.: Brookings, 1988); David Lundstrom, *A Few Good Men at UNIVAC* (Cambridge: MIT Press, 1987); James C. Worthy, *William C. Norris: Portrait of a Maverick* (Cambridge, Mass.: Ballinger, 1987).

4. See Charles Bashe et al., eds., *IBM's Early Computers* (Cambridge: MIT Press, 1986).

5. A comprehensive history of DEC is provided in Glenn Rifkin and George Harrar, *The Ultimate Entrepreneur: The Story of Ken Olsen and Digital Equipment Corporation* (Chicago: Contemporary Books, 1988).

6. On Control Data Corporation, see Worthy, *William C. Norris: Portrait of a Maverick.*

7. On the Justice Department's role, see Ernest Braun and Stuart MacDonald, *Revolution in Miniature,* 2d ed. (New York: Cambridge University Press, 1982).

8. See Braun and MacDonald, *Revolution in Miniature;* Reid, *The Chip; Electronics, Fifty Years of Electronics,* special issue, 1980; John Tilton, *International Diffusion of Technology: The Case of Semiconductors* (Washington, D.C.: Brookings, 1971); Richard Levin, "The Semiconductor Industry," in Richard Nelson, ed., *Government and Technical Progress: A Cross-Industry Analysis* (New York: Pergamon Press, 1982); Tom Forester, ed., *The Microelectronics Revolution* (Cambridge: MIT Press 1983); W.

Edward Steinmueller, "Microeconomics and Microelectronics: Economic Studies of Integrated Circuit Technology" (Ph.D. diss., Stanford University, 1987).

9. For some indication of Fairchild's innovativeness in semiconductor process technology, see Eric von Hippel, *The Sources of Innovation* (New York: Oxford University Press, 1988). According to von Hippel's research of twenty major semiconductor process innovations after 1957 (the year Fairchild was founded), five can be attributed to Fairchild.

10. Another important early semiconductor firm was Transitron, a 1950s start-up that recruited personnel from GE and Sylvania but did not reinvest in R&D and eventually failed in the mid-1960s.

11. Peter Farley, "This Biotech Alliance Focuses on Processing," *Chemical Week* (May 19, 1982):23.

12. The Kleiner of Kleiner Perkins is none other than Eugene Kleiner, one of the eight original Fairchild founders. The capital generated by the microelectronics revolution in this instance became the seed capital for another technology revolution, the biotechnology revolution. Eugene Kleiner, interview by authors, April 1988.

13. Our discussion of the organization of high-technology companies draws heavily on interviews conducted with R&D scientists, engineers, and executives during the period 1986–89. There is now a small but growing literature on the organization of high-technology companies. See especially Claudia Bird Schoonhoven and Mariann Jelinek, "Dynamic Tension in Innovative, High Technology Firms: Managing Rapid Technological Change Through Organizational Structure," in Mary Ann Von Glinow and Susan Albers Mohrman, eds., *Managing Complexity in High Technology Organizations* (New York: Oxford University Press, 1990), 90–118; Marina Gorbis, "Managing High-Technology Companies: The Art of Coping with Ambiguities" (Stanford Research Institute, SRI Business Intelligence Program, Stanford, Calif., 1986); Jay R. Galbraith, "Evolution without Revolution: Sequent Computer Systems," *Human Resources Management* 24 (Spring 1985):9–24; and Archie Kleingartner and Carolyn Anderson, eds., *Human Resources Management in High Technology Firms* (Lexington, Mass: Lexington Books, 1987).

14. Jeffrey Fox, "Biotechnology: A High-Stakes Industry in Flux," *Chemical and Engineering News* (March 29, 1982):14. For a more comprehensive discussion of labor relations in the biotechnology industry, see Martin Kenney, *Biotechnology: The University-Industrial Complex* (New Haven: Yale University Press, 1986), chap. 8.

15. Tracy Kidder, *The Soul of a New Machine* (Boston: Little, Brown, 1981), 274.

16. For a discussion of this paradox, see Frederick Brooks, *The Mythical*

Man-Month: Essays in Software Engineering (Reading, Mass.: Addison-Wesley, 1975), chaps. 1, 2.

17. David E. Lundstrom, "Small Is Beautiful," *Asian Wall Street Journal,* January 3, 1989, 4.

18. Mary Walton, "Mind behind the Mac" *Bay Area Computer Currents* (May 7, 1985):26. For further discussion of the Macintosh case, see Ikujiro Nonaka and Martin Kenney, "Innovation as an Organizational Information Creation Process: A Comparison of Canon Inc. and Apple Computer Inc.," *Hitotsubashi Business Review* (forthcoming, in Japanese).

19. As quoted in "Human Resources at Hewlett-Packard," Harvard Business School Case no. 482-125, 1982. Much the same remarks were made to us by Hewlett-Packard engineers and executives we interviewed in April 1988.

20. Interview by Martin Kenney (name withheld).

21. Intel Corporation, *A Revolution in Progress* (Santa Clara, Calif.: Intel Corporation, 1984), 44.

22. Thomas Peters and Robert Waterman, *In Search of Excellence* (New York: Warner Books, 1982), 245.

23. Quoted in Walton "Mind behind the Mac," 28–29.

24. Quoted in U.S. Congress, Joint Economic Committee, *Climate for Entrepreneurship and Innovation in the United States,* Hearings, Washington, D.C., August 30–31, 1984, 83.

25. See Nancy Dorfman, *Innovation and Market Structure: Lessons from the Computer and Semiconductor Industries* (Cambridge, Mass.: Ballinger, 1987), 7.

26. John Sculley, *Odyssey: Pepsi to Apple* (New York: Harper & Row, 1987), 133.

27. Quoted in Dirk Hanson, *The New Alchemists* (Boston: Little, Brown, 1986), 100.

28. Eugene Kleiner, interview by authors, April 1988.

29. See "Lotus Says Key Executive Has Resigned," *Wall Street Journal,* August 10, 1988, 2.

30. Gorbis, "Managing High-Technology Companies," 31.

31. See Galbraith, "Evolution without Revolution: Sequent Computer Systems," 9–24.

32. Kidder, *The Soul of a New Machine,* 64.

33. William Schroeder, president of Priam Corporation, as quoted in Gorbis, "Managing High-Technology Companies," 31.

34. See Galbraith, "Evolution without Revolution: Sequent Computer Systems," 22.

35. "In the Silicon Valley, L'Enfant Terrible Is Also L'Enfant Riche," *Wall Street Journal,* June 4, 1985, 1.

36. "Janet Axelrod," Harvard Business School Case no. 9-486-013, February 1988.

37. Quoted in Sculley, *Odyssey,* 158.

38. Jane Meredith Adams, "Valley of the Dollars," *New England Business* (February 21, 1983), 17–59.

39. Quoted in Everett Rogers and Judith Larsen, *Silicon Valley Fever: Growth of High-Technology Culture* (New York: Basic Books, 1984), 139.

40. See Barbara Berman, "The Ups and Downs of Climbing the Techno-Ladder," *Electronic Business* (March 20, 1989):68–70.

41. See Claudia Bird Schoonhoven and Kathleen Eisenhardt, *The Impact of Incubator Region on the Creation and Survival of New Semiconductor Ventures in the U.S. 1978–1986* (Report prepared for the U.S. Department of Commerce, Economic Development Administration, Washington, D.C., August 1989).

42. See Bruce Rayner and Linda Stallman, "Top 100 R&D Spenders Increase Investment by 15.7%," *Electronic Business* (January 22, 1990):90–95; Data on R&D spending by small companies as reported in "Out of the Ivory Tower," *The Economist* (February 3, 1990):65–66.

43. Quoted in U.S. Congress, Joint Economic Committee, *Climate for Entrepreneurship,* 70.

44. "Sun Microsystems Turns on the Afterburners," *Business Week,* July 18, 1988; and Rick Whiting, "Compaq Stays the Course," *Electronic Business* (October 30, 1989):25–30.

45. Quoted in U.S. Congress, Joint Economic Committee, *Climate for Entrepreneurship,* 76–77.

46. See "Hewlett-Packard: Challenging the Entrepreneurial Culture," Harvard Business School Case no. 9-384-035, 1983; "Can John Young Redesign Hewlett-Packard," *Business Week,* December 6, 1982, 72–75; "Mild Mannered Hewlett-Packard Is Making Like Superman," *Business Week,* March 7, 1988, 110–14.

47. Hanson, *New Alchemists,* 93.

48. *Digital Equipment Corporation: The First Twenty-five Years* (New York: Newcomen Society, 1983), 11.

49. Luigi Mercurio, interview by authors, April 1988.

50. Regis McKenna, quoted in U.S. Congress, Joint Economic Committee, *Climate for Entrepreneurship,* 31–34.

51. On the concept of the social structure of innovation, see Richard Florida and Martin Kenney, "Venture Capital–Financed Innovation and Technological Change in the USA," *Research Policy* 17 (1988):119–37; and Roger Miller and Marcel Cote, "Growing the Next Silicon Valley," *Harvard Business Review* 63 (July–August 1985):114–23.

52. A fascinating account is provided in Paul Freiberger and Michael

Swaine, *Fire in the Valley: The Making of the Personal Computer* (Berkeley, Calif.: Osborne/McGraw-Hill, 1984).

53. William Winter, as quoted in Braun and MacDonald, *Revolution in Miniature,* 132–33.

54. David Arscott, interview by authors, December 1986.

55. Aart DeGeus, interview by authors, April 1988. This discussion also draws upon an interview with Harvey Jones, April 1988.

56. See Schoonhoven and Eisenhardt, *The Impact of Incubator Region on the Creation and Survival of New Semiconductor Ventures in the U.S. 1978–1986.*

Chapter 4

1. Phil Kaufman, interview by authors, April 1988.

2. Burton McMurtry, interview by authors, December 1986.

3. Quoted in U.S. Congress, Joint Economic Committee, *Climate for Entrepreneurship and Innovation in the United States,* Hearings, Washington, D.C., August 27–28, 1984, 34–45.

4. The idea of "sponsored entrepreneurship" as a model of new business formation was stimulated by an interview with William Burgin of the Boston office of Bessemer Securities, June 1987.

5. Arthur Rock, "Strategy and Tactics of a Venture Capitalist," *Harvard Business Review* 65 (November–December 1987):64.

6. David Arscott, interview by authors, December 1986.

7. Interview by authors, name withheld.

8. A good topical overview of earlier forms of venture capital is provided in Thomas Doerflinger and Jack Rivkin, *Risk and Reward: Venture Capital and the Making of America's Great Industries* (New York: Random House, 1987). More detailed historical discussions are contained in Barry Supple, "A Business Elite: German-Jewish Financiers in Nineteenth-Century New York," *Business History Review* (1957):143–78; Alfred D. Chandler, Jr., "Patterns of American Railroad Finance, 1830–1850," *Business History Review* 28 (September 1954):248–63; and Glenn Porter and Harold Livesay, *Merchants and Manufacturers* (Baltimore, Md.: Johns Hopkins University Press, 1971).

9. Joseph Schumpeter, *The Theory of Economic Development* (Cambridge: Harvard University Press, 1968). See also Richard Florida and Martin Kenney, "Venture Capital–Financed Innovation and Technological Change in the USA," *Research Policy* 17 (1988):119–37.

10. An excellent history of early forms of venture capital is provided in Martha Louise Reiner, "The Transformation of Venture Capital: A History of Venture Capital Organizations in the United States" (Ph.D. diss., University of California, Berkeley, Graduate School of Business Administration, 1989).

11. See Edwin George, "Can Small Businesses Get the Capital They Need," *Duns Review* (October 1953):115–42.

12. This example is drawn from Arthur Merrill, *Investing in the Scientific Age* (Garden City, N.Y.: Doubleday, 1962).

13. Our discussion of New York venture capital is based upon oral interviews and back issues of *Venture Capital Journal*.

14. On the history of American Research and Development, see Patrick Liles, *Sustaining the Venture Capital Firm* (Cambridge: Harvard University, Management Analysis Center, 1977).

15. The Boston area had seen a number of earlier efforts to create formal venture capital institutions. During the early 1900s, the Boston Chamber of Commerce generated a small pool of risk capital and began providing managerial assistance to new enterprises. Later, in the 1930s, Boston retail magnate Edward Filene and a group of New England businessmen launched an early venture capital organization, the New England Industrial Corporation, to provide an organized form of financial and managerial assistance to new businesses. See William Leavitt Stoddard, "Small Business Wants Capital," *Harvard Business Review* 18 (Spring 1940):265–74.

16. The Committee for Economic Development's position is outlined in a series of reports that provided the background for the creation of the Small Business Administration and the later emergence of Small Business Investment Corporations. See Committee for Economic Development, *Meeting the Special Problems of Small Business* (New York: Committee for Economic Development, 1947); and A. D. H. Kaplan, *Small Business: Its Place and Problems,* Committee for Economic Development Research Study (New York: McGraw-Hill, 1948).

17. See National Association of Small Business Investment Companies (NASBIC), *Historical and Program Highlights of the SBIC Program and Private Venture Capital Investment* (Washington, D.C.: NASBIC, February 1988).

18. See Alan Ruvelson, "The First SBIC," in Stanley Rubel and Edward Novotny, eds., *How to Raise and Invest in Venture Capital* (New York: President's Publishing House, 1971), 199–208.

19. Our discussion of Chicago venture capital relies upon John W. Wilson, *The New Venturers* (San Francisco: Addison-Wesley, 1985); Gene Bylinsky, *The Innovation Millionaires* (New York: Charles Scribner's Sons, 1976); and the August 1974 and June 1981 issues of *Venture Capital Journal*.

20. Our discussion of the Silicon Valley venture capital industry draws heavily upon our interviews with the following venture capitalists in Silicon Valley and San Francisco: David Arscott, James Balderston, Frank Chambers, William Chandler, Wally Davis, Thomas Davis, Reid Den-

nis, John Dougery, William Edwards, Mary Jane Elmore, Franklin Johnson, Eugene Kleiner, Burton McMurtry, Steve Merrill, Arthur Rock, Peter Roshko, Craig Taylor, Donald Valentine, David Wegman, and Paul Wythes. Additional background is taken from back issues of *Venture Capital Journal,* which describe venture capital in California, and from a poster depicting the history of California venture capital available from Asset Management Associates, Palo Alto, California. An excellent overview of the evolution of Silicon Valley venture capital is provided in Wilson, *New Venturers.*

21. William Edwards, interview by authors, December 1986.

22. Franklin Johnson of Asset Management and Frank Chambers of Continental Capital, interviews by authors, December 1986.

23. Franklin Johnson, interview by authors, December 1986.

24. Interviews by authors with Thomas Davis, December 1986, and Arthur Rock, April 1988.

25. Our discussion of Boston venture capital draws mostly from our interviews with the following venture capitalists in the Boston area: Peter Brooke, William Burgin, Richard Burnes, Craig Burr, Charles Coulter, Thomas Claflan, Daniel Gregory, Paul Hogan, George McKinney, Joseph Powell, Patrick Sansonetti, John Shane, and Courtney Whiten. See Russell Adams, *The Boston Money Tree* (New York: Thomas Crowell, 1977), for an overview of early efforts. The early history of ARD is chronicled in Liles, *Sustaining the Venture Capital Firm.* Back issues of *Venture Capital Journal* (especially March 1974, August 1975, November 1976) describe the formation of a number of funds.

26. The 1989 figure is an estimate provided by Steven Piper of Venture Economics, Inc., January 1990.

27. Unpublished data made available to us by Horsley Keogh Associates, San Francisco, California.

28. On the falling rate of profit, see Samuel Bowles, David Gordon, and Thomas Weiskopf, "Power and Profits: The Social Structure of Accumulation and the Profitability of the Postwar U.S. Economy," *Review of Radical Political Economics* 18 (Spring–Summer 1986).

29. Though there are now many analyses of the "economic crisis" of U.S. industry, the most comprehensive and enlightening discussion of this worsening economic situation remains Michel Aglietta, *A Theory of Capitalist Regulation: The U.S. Experience* (London: New Left Books, 1979). See also Mike Davis, *Prisoners of the American Dream* (London: Verso, 1986), especially "The Political Economy of Late Imperial America."

30. Data on venture capital investment were provided by Venture Economics, Inc., January 1990. See also Richard Florida and Martin Kenney, *Venture Capital, Innovation and Economic Development* (Report to the U.S.

Department of Commerce, Economic Development Administration, Washington, D.C., September 1989).

31. Data on venture capital supply were also provided by Venture Economics, Inc., January 1990. Totals are for the states of California, Massachusetts, and New York; however the great bulk of venture capital activity in these states takes place in the regions identified here.

32. In a recent article, we make a formal distinction between venture capital that is a part of innovation complexes like Silicon Valley and Route 128 and venture capital that is located in banking centers like New York and Chicago. See Richard Florida and Martin Kenney, "Venture Capital and High-Technology Entrepreneurship," *Journal of Business Venturing* 3 (Fall 1988):301–19.

33. Florida and Kenney, *Venture Capital, Innovation and Economic Development.*

34. This is based on a database of venture capital coinvestments, which is detailed in Florida and Kenney, *Venture Capital, Innovation and Economic Development.*

35. Ibid.

36. Richard Florida and Martin Kenney, "Venture Capital, High Technology and Regional Development," *Regional Studies* 22, 1 (1988): 33–48.

37. Quoted in U.S. Congress, Joint Economic Committee, *Climate for Entrepreneurship,* 384.

38. David Morgenthaler, Sr., interview by authors, June 1985.

39. "Venture Capital Loses Its Vigor," *New York Times,* October 8, 1989.

40. John Shoch, a partner with Asset Management Corporation, Palo Alto, Calif., quoted in "Venture Capital Loses Its Vigor."

41. Quoted in "Interview with Gordon Moore: The Dark Side of Venture Capital," *Business Week,* April 18, 1983, 86.

42. John Wilson, interview by authors, March 1988.

43. See, for example, Martin Kenney, *Biotechnology: The University-Industrial Complex* (New Haven: Yale University Press, 1986).

Chapter 5

1. Quoted in U.S. Congress, Joint Economic Committee, *Climate for Entrepreneurship and Innovation in the U.S.,* Hearings, Washington, D.C., August 27–28, 1984, 80.

2. James Koford, interview by authors, April 1988.

3. See, for example, Tom Peters, *Thriving on Chaos* (New York: Knopf, 1987); and George Gilder, *Microcosm* (New York: Simon & Schuster, 1989).

4. See John Kerr, "Management's New Cry: Fight for Your Technology Rights," *Electronic Business* (August 15, 1988):44–48.

5. Interviews with high technology executives and venture capitalists in Silicon Valley and Route 128, by authors, 1987, 1988.

6. American Electronics Association, *Salary Survey of Professional Engineers* (Santa Clara, Calif.: American Electronics Association, 1986).

7. This is according to Reza Vakili of Advanced Technology Consultants. See "Competition Heats up for Qualified Workers," *Electronic Business* (December 11, 1989):107.

8. Arthur Rock, interview by authors, April 1988.

9. According to the survey, 38 percent of the CEOs answered that attracting and retaining key people was one of the biggest challenges of the next five years. As reported in Bruce Rayner, "How U.S. Electronics CEOs Will Address the New Competitive Priorities," *Electronic Business* (March 19, 1990):34–39.

10. See David Angel, "The Labor Market for Engineers in the U.S. Semiconductor Industry," *Economic Geography* 65, 2 (April 1989):99–112.

11. Angel, "Labor Market for Engineers in the U.S. Semiconductor Industry," 103.

12. See Martin Kenney, *Biotechnology: The University-Industrial Complex* (New Haven: Yale University Press, 1986), 183–84.

13. Quoted in Joel Kotkin, "The Third Wave: The Comeback in Semiconductors," *INC.* (February 1984):64.

14. Tracy Kidder, *The Soul of a New Machine* (Boston: Little, Brown, 1981), 286.

15. Quoted in John W. Wilson, *The New Venturers: Inside the High-Stakes World of Venture Capital* (Menlo Park, Calif.: Addison-Wesley, 1985), 191.

16. James Koford, interview by authors, April 1988.

17. "Seeq Technology, 1984," Harvard Business School Case Study no. 9-685-081, 1985, 5–6.

18. Source withheld.

19. Interviews with leading venture capitalists in Silicon Valley and Route 128 by authors, 1987, 1988.

20. See Robert Reich, *Tales of a New America* (New York: Times Books, 1987); and Charles Ferguson, "From the People Who Brought You Voodoo Economics," *Harvard Business Review* 66 (May–June 1988):55–62.

21. The exact figures are 101,574 total births of high-technology firms and 60,442 births of small high-technology firms, defined as those with zero to nineteen employees. These statistics are drawn from the Small Business Administration's "small business database USEEM file." Data supplied by Bruce Phillips, director of database development for the Small Business Administration's Office of Advocacy. Bruce Phillips and H. Shelton Brown, "Myths and Facts: The Role of Small High Technology

Firms in the U.S. Economy" (Unpublished report, U.S. Small Business Administration, Washington, D.C., October 1989). Data on venture capital financings supplied by Steven Piper in *Venture Economics* (January 1990). The 1,338 companies financed represent a total capital commitment of approximately $4 billion. Of this total, 401 companies, or roughly $1 billion, were first-time financings.

22. This continues: "Downstream of our 'play,' a number of service organizations make their living supporting the successes then moving their efforts to the next 'play' after each completes its 'run.' " Intel, "Silicon Valley" (Presentation to San Jose city government, September 21, 1988), 10–11.

23. On the details of Intel vs. Seeq, see Wilson, *New Venturers,* 192. Intel later filed suit against the principals of Seeq and their venture backers, and in an out-of-court settlement Seeq ultimately agreed to pay Intel an undisclosed amount and not to hire any more Intel personnel for a period of time.

24. See Wilson, *New Venturers;* and Charles Ferguson, *Technological Development, Strategic Behavior and Government Policy in Information Technology Industries* (VLSI memo no. 88-446, MIT, Cambridge, March 1988), 75–76.

25. Quoted in Kotkin, "The Third Wave: The Comeback in Semiconductors," 66. Valentine made much the same point in an interview we had with him in April 1988.

26. See "The Fairchildren: Where They Went," *San Jose Mercury News,* March 10, 1985; and, "Fairchild Semiconductor: Lily of the Valley, 1957–1987," *Electronic News* (September 28, 1987).

27. *Ibid.*

28. See "Fairchild Semiconductor: Lily of the Valley."

29. They were so highly respected that it took the venture capitalist Arthur Rock less than thirty minutes to arrange financing for this new venture. See Ernest Braun and Stuart MacDonald, *Revolution in Miniature,* 2d ed. (New York: Cambridge University Press, 1982).

30. See "Fairchild Semiconductor: Lily of the Valley." Sporck was lured by young financier Peter Sprague, who had inherited part of the Sprague Electric fortune and was trying to turn around ailing manufacturing companies.

31. Quoted in Fred Warshofsky, *The Chip War: The Battle for the World of Tomorrow* (New York: Charles Scribner's Sons, 1989), 33.

32. Ibid.

33. In the end, more than one hundred companies trace some part of their lineage back to Fairchild. Fairchild alumni like Eugene Kleiner and Donald Valentine also played extremely active roles in the evolution of Silicon Valley venture capital.

34. On the demise of Fairchild, see David Chagall, "How the Chip Lobby Short-Circuited Fujitsu," *California Business* (July 1987):40–51; "Bottom Line Dictated Who Would Buy Fairchild," *San Jose Mercury News,* September 7, 1987; "Fairchild Flounders under Schlumberger," *San Jose Mercury News,* March 10, 1985; "Fairchild to Pursue Close Ties with Fujitsu Anyway," *San Jose Mercury News,* March 18, 1987. It is interesting to note that National then laid off thousands of Fairchild employees. The company was saved from Fujitsu at the cost of thousands of jobs.

35. See Glenn Rifkin and George Harrar, *The Ultimate Entrepreneur: The Story of Ken Olsen and Digital Equipment Corporation* (Chicago: Contemporary Books, 1988), 95–100.

36. See James C. Worthy, *William C. Norris: Portrait of a Maverick* (Cambridge, Mass.: Ballinger, 1987), 44–45; and "Cray Research, Inc.," Harvard Business School Case no. 385-011, 1984.

37. See "How Steve Chen Plans to Beat Seymour Cray at His Own Game," *Electronics* (March 3, 1988); Mr. Mitchell Waldrop, "Cray Supercomputer Axed, Superstar Departs," *Science* 237 (September 25, 1987):1558–59; "Survival of Cray Spin-off Will Depend on Creativity and Longevity of Designer," *Wall Street Journal,* November 2, 1989; Michael Leibowitz, "Clash of the High-Speed Titans," *High Technology Business* (July 1988):48–62; and "Supercomputers: The Proliferation Begins," *Electronics* (March 3, 1988):51–77.

38. Convex was launched in 1982 by Robert Pauluck and Steven Wallach. The latter was a member of the Eagle team immortalized in Kidder, *The Soul of a New Machine.* Alliant was also launched in 1982 by Craig Mundie from Data General's North Carolina development group that competed against the Eagle team. See First Boston Equity Research, "The Minisupers: Alliant Computer Systems and Convex Computer" (June 30, 1987); Dataquest, "The Minisupercomputer Market Grows Up: The Prime/Cydrome Entry," *Dataquest Research Newsletter* (January 1988); Tony Greene, "Mentor Graphics Wants to Rebuild the Design Industry," *Electronic Business* (February 5, 1990):20–23.

39. See Kidder, *The Soul of a New Machine.*

40. See Hewlett-Packard's in-house magazine, *Measure,* "Our Entrepreneurial Alumni" (July–August 1984):16–19.

41. Daniel Okimoto, Takuo Sugano, and Franklin Weinstein, eds., *Competitive Edge: The Semiconductor Industry in the U.S. and Japan* (Stanford, Calif.: Stanford University Press, 1984), 188–89.

42. See John Sculley, *Odyssey: Pepsi to Apple* (New York: Harper & Row, 1987); and Frank Rose, *West of Eden: The End of Innocence at Apple Computer* (New York: Viking, 1989). See also "Apple Computer Tries to Achieve Stability but Remain Creative," *Wall Street Journal,* July 16, 1987, 1.

43. "Seeq Technology, 1984," Harvard Business School Case Study no. 9-685-081, 1985, 5–6. An economics perspective on the issues regarding human resource investment in innovative firms is provided. Wesley Cohen and Daniel Levinthal, "Fortune Favors the Prepared Firm" (Unpublished manuscript, Carnegie Mellon University, Pittsburgh, Pa.)

44. Quoted in Everett Rogers and Judith Larsen, *Silicon Valley Fever* (New York: Basic Books, 1984), 153.

45. See Joseph Schumpeter, *Capitalism, Socialism and Democracy* (New York: Harper & Row, 1975). Much the same point is also advanced by Abernathy and Utterback in their classic paper on the product cycle and technological innovation; see William Abernathy and James Utterback, "Patterns of Industrial Innovation," *Technology Review* 80 (1978).

46. As quoted in Peter Burrows, "Industry Pioneers Offer Guideposts for the Future," *Electronic Business* (December 11, 1989):21. From an economics perspective, this could be seen as a special form of "knowledge spill-over" that undercuts a firm's incentive to invest in innovation. See Wesley Cohen and Richard Levin, "Empirical Studies of Innovation and Market Structure," in R. Schmalensee and R. Willig, eds., *Handbook of Industrial Organization* 2 (Amsterdam: North-Holland, 1989):1059–1107.

47. Gordon Moore, "Entrepreneurship and Innovation: The Electronics Industry," in Ralph Landau and Nathan Rosenberg, eds., *The Positive Sum Strategy: Harnessing Technology for Economic Growth* (Washington, D.C.: National Academy Press, 1986), 424.

48. Phil Kaufman, interview by authors, April 1988.

49. See Dataquest, "Aggressive Startups Spawn New Companies," *Dataquest Research Bulletin* (June 1988).

Chapter 6

1. "Interview with Gordon Moore: The Dark Side of Venture Capital," *Business Week,* April 18, 1983, 86.

2. Silicon Valley and Route 128 are clearly the nation's dominant centers of high technology. In 1988, twenty-seven of the fastest-growing U.S. electronics companies were located in Silicon Valley; eleven were in Boston's Route 128 area; and eleven were in Los Angeles. Outside of California and Massachusetts, no other state had more than six. See "1988's Fastest Growing U.S. Companies," *Electronic Business* (May 1, 1988):36–38.

3. See George Gilder, *Microcosm: The Quantum Revolution in Economics and Technology* (New York: Simon & Schuster, 1989).

4. George Gilder, "The Law of the Microcosm" *Harvard Business Review* (March–April 1988):55. In an article published in *Electronic Business,*

Gilder argues further: "The new technologies of the microcosm—artificial intelligence, silicon compilation, and parallel processing—all favor entrepreneurs and small companies. All three enable entrepreneurs to use the power of knowledge to economize on capital and enhance its efficiency, mixing sand and ideas to generate wealth and power. See Gilder, "The New American Challenge," *Electronic Business* (August 7, 1989):40.

5. Sabel goes further in a recent article, arguing that flexible small-firm districts are self-adaptive systems that grow and learn. See Sabel, "Flexible Specialization and the Re-emergence of Regional Economies," in P. Hirst and J. Zeitlin, eds., *Reversing Industrial Decline* (New York: St. Martin's Press, 1989), 17–70. A large body of work on the organization of Silicon Valley has been produced by Allen Scott and Michael Storper. See Scott and Storper, "High Technology Industry and Regional Development: A Critique and Reconstruction," *International Social Science Review* 112 (May 1987):215–32; and Storper and Scott, "The Geographical Foundations and Social Regulation of Flexible Production Complexes," in Jennifer Wolch and Michael Dear, eds., *The Power of Geography* (Boston: Allen & Unwin, 1988).

6. See AnnaLee Saxenian, "The Political Economy of Industrial Adaptation in Silicon Valley" (Ph.D. diss., MIT, 1988), chap. 5, p. 17.

7. Thomas Howell, William Noellert, Janet MacLaughlin, and Alan Wolff, *The Microelectronics Race: The Impact of Government Policy on International Competition* (Boulder, Colo.: Westview Press, 1988), 12.

8. Dwight Davis, "Making the Most of Your Vendor Relationships," *Electronic Business* (July 10, 1989):42.

9. Sylvia Tiersten, "The Changing Face of Purchasing," *Electronic Business* (March 20, 1989):22.

10. As quoted in Davis, "Making the Most of Your Vendor Relationships," 44–45.

11. See David Burt, "Managing Suppliers Up to Speed," *Harvard Business Review* (July–August 1989):127–135.

12. See Richard Gordon, "Reseaux globaux et le processus d' innovation dans les petites et moyennes enterprises: le cas de Silicon Valley," forthcoming in D. Malliat and J.C. Perrin, eds., *Enterprises Innovatrices et Reseaux Locaux* (Paris: ERESA-Economica, 1990).

13. See Alden Hayashi, "The New Shell Game: Where Was This U.S. Chip Really Made," *Electronic Business* (March 1, 1988):36–40.

14. See Gilder, *Microcosm;* and Saxenian, "Political Economy of Industrial Adaptation in Silicon Valley."

15. See Jim Jubak, "Venture Capital and the Older Company," *Venture* (September 1988):44–45.

16. Rick Whiting, "Exploring Sun's Solar System," *Electronic Business* (January 3, 1989):46–50.

17. Name withheld, Fujitsu America Inc. executive, interview by Martin Kenney, December 1989.
18. As quoted in "The Fujitsu Difference," special advertising section, *Business Week,* December 25, 1989, 91.
19. Quoted in Davis, "Making the Most of Your Vendor Relationships," 43.
20. Shillman's comments are from "Globalization Is the Driving Force," *Electronic Business* (March 19, 1990):44–45.
21. Gazelle's comments are from Gene Bylinsky, "The Hottest High-Tech Company in Japan," *Fortune,* January 1, 1990, 83–88.
22. As quoted in Edward Welles, "The Tokyo Connection," *INC.* (February 1990):59–64.
23. See Valerie Rice, "What's Right with America's IC Equipment Makers?" *Electronic Business* (May 15, 1989):28–44; "Key Technology Might Be Sold to the Japanese," *New York Times,* November 27, 1989, 1.
24. Jay Stowsky, "Weak Links, Strong Bonds: U.S.-Japanese Competition in Semiconductor Production Equipment," in Chalmers Johnson, Laura Tyson, and John Zysman, eds., *Politics and Productivity* (Cambridge, Mass.: Ballinger, 1989), 241–74.
25. See John Kerr, "Breaking into Japan: Small U.S. Companies Show How It's Done," *Electronic Business* (November 13, 1989):72–76.
26. Rick Whiting, "MRS Exports Technology to Japan," *Electronic Business* (March 19, 1990):105–8.
27. Interviews by authors with Phil Kaufman of Silicon Compiler Systems, Kamran Elahian of Cirrus Logic, Aart de Geus and Harvey Jones of Synopsis, James Solomon of SDA, and Kenneth Levy of KLA Instruments.
28. See Valerie Rice, "Breaking into Japan: Small U.S. Companies Find Success in a Demanding Market," *Electronic Business* (November 27, 1989):60–62.
29. Linda Stallman and Bruce Rayner, "Vibrant Foreign Sales Can't Mask Discouraging Indicators," *Electronic Business* (January 22, 1990).
30. Harvey Jones, interview by authors, April 1988.
31. As quoted in John W. Wilson, *The New Venturers: Inside the High-Stakes World of Venture Capital* (Menlo Park, Calif.: Addison-Wesley, 1985), 195. Valentine offered much the same assessment in our interview with him.
32. See Wilson, *New Venturers,* 195.
33. See William Sahlman and Howard Stevenson, "Capital Market Myopia," *Journal of Business Venturing* 1 (Winter 1985):7–14. See also "The Disk Drive Maker That's Driving to a Record," *Business Week,* September 14, 1987, 134–36; "Seagate Goes East and Comes Back a Winner," *Business Week,* March 16, 1987, 94; John McCreadie, "Vola-

tile Hard Drive Arena Picking Up the Pieces," *Electronic Business* (November 27, 1989):65–68; and "Conner's Peripheral Vision is Still Sharp," *Wall Street Journal,* April 13, 1988, 28.

34. See, "After a Domestic Shakeout, U.S. Drive Makers Face Japan," *New York Times,* January 21, 1990.

35. See Martin Kenney, *Biotechnology: The University-Industrial Complex* (New Haven: Yale University Press, 1986).

36. Luigi Mercurio, interview by authors, April 1988.

37. "The Look of the Industry in 2000," *Electronics* (April 2, 1987).

38. *The Economist,* June 11, 1988, 65.

39. Our discussion of the mainframe segment is drawn from Dataquest and Standard and Poor's Computer and Office Equipment Industry Survey. A good overview of recent trends appears in "A Bold Move In Mainframes: IBM Plans to Make Them Key to Networking and So Restore Its Growth," *Business Week,* May 29, 1989, 72–78; "Who's Ahead in the Computer Wars?" *Fortune,* February 2, 1990, 59–66.

40. An overview of the minicomputer industry is provided in Elaine Romanelli, "New Venture Strategies in the Minicomputer Industry," *California Management Review* (Fall 1987):160–75.

41. For further discussion, see Jeffrey Bairstow, "Personal Workstations Redefine Desktop Computing" *High Technology* (March 1987); and Dataquest, "A Comparative Look Inside America's Leading 32-Bit PCs," *Dataquest Research Newsletter* (March 1988). On Sun, see "High Noon for Sun," *Business Week,* July 24, 1989, 70–75.

42. A good review of developments can be found in Rick Whiting, "Personal Computer Have-Nots Fight for Bigger Slice of Market," *Electronic Business* (October 30, 1989):34–35.

43. Peter Burrows, "Power Packed Portables Are Storming PC Market," *Electronic Business* (January 23, 1990):70–73; see also "The Big News in Tiny Computers," *New York Times,* May 14, 1989.

44. "Supercomputers: The Proliferation Begins," *Electronics* (March 3, 1988):51–77.

45. See "Cray Research, Inc.," Harvard Business School Case no. 385-011, 1984; and James C. Worthy, *William C. Norris: Portrait of a Maverick* (Cambridge, Mass.: Ballinger, 1987).

46. See J. Michael Ruhl, "Cray under Fire," *Electronic Business* (February 20, 1989):18–47; "How Steve Chen Plans to Beat Seymour Cray at His Own Game," *Electronics* (March 3, 1988):54; M. Mitchell Waldrop, "Cray Supercomputer Axed, Superstar Departs," *Science* 237 (September 25, 1987):1558–59.

47. See First Boston Equity Research, "The Minisupers: Alliant Computer Systems and Convex Computer" (June 30, 1987); Dataquest, "Ardent Computer Introduces a New Class of Computers: Graphics Supercomputers," *Dataquest Research Newsletter* (March 1988); Dataquest, "The

Minisupercomputer Market Grows Up: The Prime/Cydrome Entry,"
Dataquest Research Newsletter (January 1988).
48. See "Supercomputers: the Proliferation Begins," *Electronics* (March 3,
1988):51–77; Parallel Processing," *Electronic Business* (February 15,
1987):28–30; Kathleen Wiegner, "Chain Gang Computing," *Forbes,*
April 2, 1990, 154–162; Clifford Isberg, "A Taxonomy of Parallel
Processors" (Palo Alto, Calif.: Stanford Research Institute, 1988).
49. See C. Gordon Bell et al., "Supercomputing for One," *IEEE Spectrum*
(April 1988):46–50; and Michael Leibowitz, "Workstation Wars: The
Battle of the Big 7," *High Technology Business* (November 1987):22–29.
50. Kenneth Flamm, *Creating the Computer* (Washington, D.C.: Brookings,
1988).
51. See Rick Whiting, "Whither IBM: Can the Giant Reassert Itself?" *Electronic Business* (December 11, 1989):112–14.
52. See John McCreadie and Valerie Rice, "Nine New Mavericks: The
Semiconductor Fraternity of the '90's," *Electronic Business* (September 4,
1989):30–36; and Valerie Rice, "Where Are They Now: 1987's Superstars Revisited," *Electronic Business* (September 4, 1989):36–38. Our
discussion of the ASIC industry draws from personal interviews with
David Fullager of Maxim Integrated Products and James Koford of LSI
Logic, April 1988; Kamran Elahian of Cirrus Logic, April 1988; and
James Koford of LSI Logic, April 1988.
53. See C. Gordon Bell, "A Surge for Solid State," *IEEE Spectrum* (April
1986):73.
54. See Gilder, *Microcosm;* and T. J. Rogers, "Landmark Messages from the
Microcosm," *Harvard Business Review* (January–February 1990):25–30.
55. See Dataquest, "IBM: Semiconductor Supplier and Buyer," *Dataquest
Research Newsletter* (January–March 1988).
56. See "Intel: Supplier Rising as a Big Competitor," *New York Times,*
February 14, 1990; and Robert Ristelhueber, "Chip-Makers Seek to
Emulate 386, 486," *Electronic News* (March 12, 1990):1, 20.
57. On vertical integration, see Alfred Chandler, Jr., *Strategy and Structure:
Chapters in the History of Industrial Enterprise* (Cambridge: MIT Press,
1962); and Oliver Williamson, *Markets and Hierarchies: A Study in the
Economics of Internal Organization* (New York: Free Press, 1975): This
model of industrial development via integration is also reflected in product cycle theories of industrial development. See Raymond Vernon,
"International Investment and International Trade in the Product
Cycle," *Quarterly Journal of Economics* 80 (1966); idem, *Sovereignty at Bay*
(New York: Basic Books, 1971); and Robert Gilpin, *U.S. Power and the
Multinational Corporation* (New York: Basic Books, 1975).
58. Robert Noyce, "Competition and Cooperation: A Prescription for the
Eighties," *Research Management* (March 1982):13–17.
59. For an overview of the difficulties of the "Route 128 miracle" and the

broader "Massachusetts miracle" of which it was a part, see Bennett Harrison and Jean Kluver, "Re-assessing the 'Massachusetts Miracle': Reindustrialization and Balanced Growth, or Convergence to Manhattanization," *Environment and Planning A* 21 (1989):771–801.

60. See Fumio Kodama, "Japanese Innovation in Mechatronics Technology," *Science and Public Policy* 13 (February 1988); Kansai Productivity Center, *Mechatronics: The Policy Ramifications* (Tokyo: Asian Productivity Association, 1985). For a more general discussion of innovation and technology fusion, see Abbott Usher, *A History of Mechanical Inventions* (Cambridge: Harvard University Press, 1954).

61. For an excellent discussion of systems innovations, see Thomas Hughes, *Networks of Power* (Baltimore, Md.: Johns Hopkins University Press, 1983), chaps. 1, 2.

62. See George Gamota and Wendy Frieman, *Gaining Ground: Japan's Strides in Science and Technology* (Cambridge, Mass.: Ballinger, 1988).

63. On the U.S. superconductor effort, see U.S. Congress, Office of Technology Assessment, *Commercializing High-Temperature Superconductivity* (Washington, D.C.: U.S. Government Printing Office, 1988); Simon Foner and Terry Orlando, "Superconductors: The Long Road Ahead," *Technology Review* (February–March 1988):46–47; David Gumpert, "The Big Chill," *Electronic Business* (January 23, 1989):24–32; Robert Haavind, "HDTV: Who Will Win the Battle to Set U.S. Standards," *Electronic Business* (January 8, 1990):24–25.

64. See SRI International, "The U.S. Home Entertainment Electronics Market" (Report no. 749, SRI International, Business Intelligence Program, Stanford, Calif., 1987).

65. George Morrow, "U.S. Industry Is Setting Itself Up to Become Junior Player in DRAM Market" *Info World* (August 1, 1988):36.

66. Hugh Carter Donahue, "Choosing TV of the Future," *Technology Review* (April 1989):31–40.

67. As quoted in "Supertelevision: The High Promise and High Risk of High-Definition TV," *Business Week,* January 30, 1989, 57.

68. See Regis McKenna, "Large-Small Company Alliances Can Fuel Growth," *Electronic Business* (November 13, 1989):93–96.

69. An important exception is Motorola, which produces pocket pagers and car phones that use integrated circuits.

70. See Semiconductor Industry Association, *Meeting the Global Challenge: Advanced Electronics Technology and the Semiconductor Industry* (Cupertino, Calif.: Semiconductor Industry Association, 1989). The Semiconductor Industry Association estimates that foreign integrated electronics firms consume up to 35 percent of their semiconductor production in-house.

71. Semiconductor Industry Association, *Meeting the Global Challenge,* 12.

72. See "U.S. Aid Sought for Electronics," *New York Times,* October 3, 1989, 1.

73. For an absolutely fascinating perspective on the continuing difference between the work environments in high-technology and traditional industrial corporations (which is drawn from the comments of newly minted engineers and M.B.A.s), see Barbara Bergman "The Best and the Brightest Speak Out," *Electronic Business* (June 12, 1989):29–37.
74. See Jeffrey Zygmont, "Silicon Valley and Detroit Shuck off Past Animosities," *Electronic Business* (April 3, 1989):100–104.
75. Ibid., 102.
76. See Richard Florida, Martin Kenney, and Andrew Mair, "The Transplant Phenomena: Japanese Automobile Manufacturers in the United States," *Economic Development Commentary* 12 (Winter 1989):3–9.

Chapter 7

1. Valerie Rice and Carol Suby, "Sematech: United We Stand," *Electronic Business* (May 1, 1988):32. Chips and Technologies is a successful recent semiconductor start-up.
2. Quoted in "Back to the Basics," *Business Week,* 1989 special issue, "Innovation in America," 15.
3. For a discussion of this particular method of conceptualizing innovation, see Michael Storper and Richard Walker, *The Capitalist Imperative* (London: Basil Blackwell, 1989).
4. Rodney Smith, quoted in "A Look at the Industry in the Year 2000," *Electronics* (April 2, 1987):65.
5. Anthony Friscia, president of Advanced Manufacturing Research, quoted in Dwight Davis, "Beating the Clock," *Electronic Business* (May 29, 1989):21–28.
6. These data are from the U.S. Equal Employment Opportunity Commission and include high-tech electronics, software, and high-tech–related business services as reported in *Global Electronics* (February 1990). In 1988, there were 192,665 total employees of Silicon valley high-tech firms, including 69,645 professionals, 29,312 officials and managers, 21,257 technicians, 4,472 sales, 25,335 office and clerical, 8,290 craft, 27,555 operatives, 1,528 laborers, and 5,271 service workers. These data slightly underestimate total employment since they are confined to employers with 100 or more workers. For example, the high-tech electronics sectors that reported a total of 122,000 workers to the EEOC actually employed 195,000 workers in 1988.
7. Investment data as reported in "Economy Drifts Down, Electronic Markets Follow," *Electronic Business* (January 8, 1990):60–61.
8. State of California, *Annual Planning Information 1988–89 for San Jose Standard Metropolitan Statistical Area* (California Employment Development Department, May 1988), 39.
9. See "Silicon Valley's Workforce Remains Segregated," *Global Electronics*

(February 1990):1–4; see also Lenny Siegel and Herb Borock, *Background Report on Silicon Valley*, Report to the U.S. Civil Rights Commission (Mountain View, Calif.: Pacific Studies Center, 1982).

10. Steve Early and Rand Wilson, "Do Unions Have a Future in High Technology?" *Technology Review* (October 1986):61.

11. For an excellent study of management-labor relations inside a high-technology firm in Silicon Valley, see Soon Kyoung Cho, "The Labor Process and Capital Mobility: The Limits of the New International Division of Labor," *Politics and Society* 14, 2 (1985):185–222.

12. State of California, *Annual Planning Information 1988–89 for San Jose Standard Metropolitan Statistical Area*, 39.

13. Bruce Rayner and Linda Stallman, "Top Bosses' Pay Slows in Sluggish Electronics Sector," *Electronic Business* (September 18, 1989):60–66.

14. "Apple's Finance Chief Got $1.5 Million Signing Bonus," *Wall Street Journal*, January 4, 1990, B4.

15. Judith Stacey, "Sexism by a Subtler Name," *Socialist Review* 90 (1987). See also two earlier studies: Cho, "Labor Process and Capital Mobility"; and Seigel and Borock, *Background Report on Silicon Valley*.

16. On environmental pollution problems in high-technology manufacturing, see Lenny Siegel and John Markoff, *The High Cost of High Tech* (New York: Harper & Row, 1985).

17. See Joseph LaDou, "The Not-So-Clean Business of Making Chips," *Technology Review* (May–June 1984):23–36. See also the recent issue of *Electronic Business* (September 18, 1989) devoted to the "toxic troubles" of the semiconductor industry.

18. See Philip Shapira, "Industry and Jobs in Transition: A Study of Industrial Restructuring and Worker Displacement in California" (Ph.D. diss., University of California, Berkeley, 1986).

19. See Siegel and Borock, *Background Report on Silicon Valley*, 40–42.

20. Everett Rogers and Judith Larsen, *Silicon Valley Fever* (New York: Basic Books, 1984), 150.

21. John Wilke, "Firms Oust 'No-Layoff' Tradition," *Wall Street Journal*, April 13, 1990, B2.

22. See Richard Walker, "The Playground of U.S. Capitalism? The Political Economy of the San Francisco Bay Area," *The Year Left* 4 (1990):3–80. See also "More Companies Rent Temporary Workers to Fill Gaps," *San Jose Mercury News*, September 28, 1986.

23. Everett Kassalow, "The Unions' Stake in High-Tech Development," in Archibald Kleingartner and Carolyn Anderson, eds., *Human Resource Management in High Technology Firms* (Lexington, Mass.: Lexington Books, 1987), 157–82.

24. Early and Wilson, "Do Unions Have a Future in High Technology?"; Siegel and Markoff, *The High Cost of High Tech*.

25. See Cho, "Labor Process and Capital Mobility."
26. Early and Wilson, "Do Unions Have a Future in High Technology?" 61.
27. Ibid., 61–62.
28. See Kassalow, "The Unions' Stake in High-Tech Development."
29. On this suit, see Early and Wilson, "Do Unions Have a Future in High Technology?" 62–63.
30. Allen Scott, "The Semiconductor Industry in Southeast Asia: Organization, Location, and the International Division of Labor," *Regional Studies* 21, 2 (1987):143–59. See John Alic and Martha Caldwell Harris, "Employment Lessons from the Electronics Industry," *Monthly Labor Review* (February 1986):27–36.
31. Robin Agarwal, "Motorola Bets Big on the China Card in Hong Kong," *Electronic Business* (March 5, 1990):53–54.
32. Dataquest, unpublished data, March 1988.
33. On the IBM personal computer, see "America's High-Tech Crisis: Why Silicon Valley Is Losing Its Edge," *Business Week,* March 11, 1985; "Seagate Goes East—and Comes Back a Winner," *Business Week,* March 16, 1987, 94. To ensure continued communication and good quality control, Seagate brought over fifty foreign workers to Seagate's headquarters to teach them production techniques. This strategy won Seagate fifty percent of the 5.25 inch disk drive market, but now Seagate seems to have slipped in new product development. See Alden Hayashi, "Hard Times for Hard Drives," *Electronic Business* (November 15, 1988):32–37.
34. Ernst & Young, *Electronics 90—The New Competitive Priorities: A Survey of Chief Executive Officers in the U.S. Electronics Industry* (San Francisco: Ernst & Young, 1990), 73.
35. "America's High-Tech Crisis," 57.
36. Constantinos Markides and Norman Berg, "Manufacturing Offshore Is Bad Business," *Harvard Business Review* 66 (September–October 1988):113–20.
37. See Intel Corporation, *A Revolution in Progress* (Santa Clara, Calif.: Intel Corporation, 1984).
38. See Alic and Harris, "Employment Lessons from the Electronics Industry."
39. Doreen Massey, *Spatial Divisions of Labor: Social Structures and the Geography of Production* (New York: Methuen, 1984). On the broader phenomenon of the "new international division of labor," see Folker Froebel, Jurgen Heinrichs, and Otto Kreye, *The New International Division of Labor* (New York: Cambridge University Press, 1980).
40. Donald Valentine, interview by authors, April 1987.
41. In addition, only 40 percent of the employees of U.S. semiconductor

firms were engaged in production work, compared with more than 80 percent of the workers in overseas plants. There is good reason to believe these differences have increased since then. See Alic and Harris, "Employment Lessons from the Electronics Industry."

42. Philip Liu, "Big Blue Benefits from R&D in Taiwan," *Electronic Business* (May 15, 1989):105–6.

43. "Is the Era of Cheap Asian Labor Over?" *Business Week,* May 15, 1989, 45–46.

44. Alden Hayashi, "The New Shell Game: Where Was This U.S. Chip Really Made," *Electronic Business* (March 1, 1988):36.

45. Dataquest, "The Electronics Migration to Korea: Who and Why," *Dataquest Research Newsletter* (June 1988).

46. Ernst & Young, *Electronics 90—The New Competitive Priorities,* 74.

47. Hayashi, "The New Shell Game," 36–40.

48. See Bruce Rayner, "The Amazing Shrinking Flexible Fab of the Future," *Electronic Business* (September 18, 1989):17–18.

49. As quoted in Hayashi, "The New Shell Game," 40.

50. "Intel: The Next Revolution," *Business Week,* September 26, 1988, 78–79.

51. Hayashi, "The New Shell Game," 37.

52. On hollowing, see Norman Jonas, "The Hollow Corporation," *Business Week,* March 3, 1986, 57–85. On deindustrialization, see Barry Bluestone and Bennett Harrison, *The Deindustrialization of America: Plant Closings, Community Abandonment, and the Dismantling of Basic Industry* (New York: Basic Books, 1982).

53. See, for example, Andrew Sayer and Kevin Morgan, "High Technology Industry and the International Division of Labour: The Case of Electronics," in Michael Breheny and Ronald McQuaid, eds., *The Development of High Technology Industries: An International Survey* (London: Croom Helm, 1987).

54. As quoted in Jonas, "The Hollow Corporation," 75. See also Akio Morita, "Something Basic Is Wrong in America, *New York Times,* October 1, 1989.

55. Hambrecht and Quist data cited in Jeffery Bairstow, "Can the U.S. Semiconductor Industry Be Saved?" *High Technology* (May 1987):34.

56. See, Dataquest, "Wafer Fab Update: A Tour of the New North American Leading Edge Fabs," *Dataquest Research Newsletter* (May 1988).

57. James Koford, interview by authors, April 1988. See also Stephen Cohen and John Zysman, *Manufacturing Matters* (New York: Basic Books, 1987).

58. Markides and Berg, "Manufacturing Offshore Is Bad Business," 117.

59. Quoted in Hayashi, "The New Shell Game," 38.

60. On Korea, see Alice Amsden, *Asia's Next Giant: South Korea and Late Industrialization* (New York: Oxford University Press, 1989).

61. See Dataquest, "South Korean Semiconductor Industry Shoots for the Moon in 1988," *Dataquest Research Newsletter* (January 1988). On Hyundai, see Alden Hayashi, "Hyundai's Headaches," *Electronic Business* (February 6, 1989):25–32.

62. "Intel Ensuring Chip Supply Allies with Japanese Firm," *Wall Street Journal,* January 13, 1990, B8; and "Advanced Micro Devices to Swap Factory for Sony Expertise," *Wall Street Journal,* February 21, 1990, B4.

63. This term was coined by Tessa Morris-Suzuki, "Robots and Capitalism," *New Left Review* 147 (1984):109–21. See also Morris-Suzuki, *Beyond Computopia: Information, Automation and Democracy in Japan* (London: Routledge Kegan Paul International, 1988). For a later adoption of this term, see Don Kash, *Perpetual Innovation: The New World of Competition* (New York: Basic Books, 1989).

Chapter 8

1. James Solomon, interview by authors, April 1988. SDA (now Cadence) is a leading Silicon Valley computer-aided engineering start-up.

2. See Chalmers Johnson, *MITI and the Japanese Miracle* (Stanford, Calif.: Stanford University Press, 1982). A more recent and somewhat more balanced account is Marie Anchordoguy, *Computers, Inc: Japan's Challenge to IBM* (Cambridge: Harvard Council on East Asian Studies, 1989). A popular version of this view is reported in a *Business Week* cover story entitled "Rethinking Japan," August 7, 1989, 44–52.

3. See Charles Ferguson, "From the People Who Brought You Voodoo Economics," *Harvard Business Review* 66 (May–June 1988):55–62.

4. See Knuth Dohse, Ulrich Jurgens, and Thomas Malsch, "From 'Fordism' to 'Toyotism'? The Social Organization of the Labor Process in the Japanese Automobile Industry," *Politics and Society* 14, 2 (1985):115–46. See also Mike Parker and Jane Slaughter, "Management by Stress," *Technology Review* 91 (October 1988):36–44.

5. See Kim Clark, W. Bruce Chew, and Takahiro Fujimoto, "Product Development in the World Auto Industry," *Brookings Papers on Economic Activity* 3 (1987); and Kim Clark and Takahiro Fujimoto, "Overlapping Problem Solving in Product Development" (Working paper, Harvard Business School, Cambridge, March 1987). See also Edwin Mansfield, "The Speed and Cost of Industrial Innovation in Japan and the United States: External vs. Internal Technology," *Management Science* (October 1988):1157–68.

6. Moriya Uchida, quoted in "A Wave of Ideas, Drop by Drop," *Business Week,* 1989 special issue, "Innovation in America."

7. The literature on Japanese firms is immense. For a very readable discussion, see James Abegglen and George Stalk, *Kaisha: The Japanese Corporation* (New York: Basic Books, 1985).

8. See Gellman Research Associates, *Indicators of International Trends in Technological Innovation* (Washington, D.C.: Gellman Research Associates, 1976). Though the study is somewhat dated, our own perusal of patent and innovation statistics in the two countries gives us little reason to believe that the relative balance has changed substantially in the last decade.

9. Alden Hayashi, "NEC Takes the Triple Crown in Electronics," *Electronic Business* (September 15, 1987):40–48. See also Arthur D. Little Decision Resources, "The 20 Leading Competitors in the World Information Services Industry, 1987 and 1992," *Arthur D. Little Spectrum: Information Systems and Technologies Overview* (February 1988).

10. See Gene Gregory, *Japanese Electronics Technology: Enterprise and Innovation* (Tokyo: Japan Times, 1986); Daniel Okimoto, Takuo Sugano, and Franklin Weinstein, *Competitive Edge: The Semiconductor in the US and Japan* (Stanford, Calif.: Stanford University Press, 1984); W. Edward Steinmueller, "Industry Structure and Government Policies in the U.S. and Japanese Integrated Circuit Industries" (Center for Economic Policy Research Working Paper no. 105, Stanford University, Stanford, Calif., December 1986); and U.S. Congress, Office of Technology Assessment, *International Competitiveness in Electronics* (Washington, D.C.: U.S. Government Printing Office, 1983).

11. These figures are adapted from Dodwell Marketing Consultants, *Key Players in Japanese Electronics* (Tokyo: Dodwell Marketing Consultants, 1985).

12. Research manager of a large Japanese electronics firm, name withheld, interview by Martin Kenney, June 1988.

13. Hitachi Corporation, "Ten Percent Memorandum" (1984).

14. See Semiconductor Industry Association, *Meeting the Global Challenge: Advanced Electronics Technology and the American Semiconductor Industry* (Cupertino, Calif.: Semiconductor Industry Association, 1989), 7.

15. "Japanese Ingenuity," *Wall Street Journal,* October 25, 1989.

16. Gene Gregory, "Mega-research Investment for Japanese Microelectronics," *Research Management* (May–June 1983):15.

17. Quoted in Arthur Klausner, "Today's Trends," *Bio/Technology* 5 (October 1988):1019–26.

18. Ronald Rohrer of Carnegie Mellon University, personal communication with Richard Florida, November 1987.

19. See Fumio Kodama, "Japanese Innovation in Mechatronics Technology," *Science and Public Policy* 13 (February 1988).

20. See the chapter entitled "Mechatronics" in George Gamota and Wendy Frieman, *Gaining Ground: Japan's Strides in Science and Technology* (Cambridge, Mass.: Ballinger, 1988), 85–106; and James Nevins, "Mechatronics," in Cecil Uyehara, ed., *U.S.-Japan Science and Technology Ex-*

change: Patterns of Interdependence (Boulder, Colo.: Westview Press, 1988), 92–142.

21. See Gamota and Frieman, *Gaining Ground,* 41–64; and Jonathan Joseph, "How the Japanese Became a Power in Optoelectronics," *Electronics* (March 17, 1986):50–51.

22. " 'Photonics': The U.S. Is Losing Ground to You-Know-Who," *Business Week,* September 19, 1988, 88.

23. See Ken-ichi Imai, "Network Industrial Organization and Incremental Innovation in Japan" (Discussion Paper no. 122, Institute of Business Research, Hitotsubashi University, Tokyo, May 1988); and idem, "Japanese Corporate Strategies toward International Networking and Product Development" (Unpublished paper, Institute of Business Research, Hitotsubashi University, Tokyo, October 1988).

24. See Ronald Dore, "Goodwill and the Spirit of Market Capitalism," *British Journal of Sociology* 34, 4 (1983):459–82. For further discussion of the concept of obligational subcontracting, see Ronald Dore, *Taking Japan Seriously* (Stanford, Calif.: Stanford University Press, 1987), 169–92.

25. Our position is developed in greater detail in Richard Florida and Martin Kenney, "High Technology Restructuring in the USA and Japan," *Environment and Planning A* 22 (1990):233–52.

26. Michiyoshi Hagino, deputy general manager, Tochigi Research Facility, Honda R&D Ltd. Co., interview by Martin Kenney, June 1988.

27. Masahiko Aoki, *Information, Incentive and Bargaining in the Japanese Economy* (Cambridge: Cambridge University Press, 1989).

28. See Masahiko Aoki, "The Japanese Firm in Transition," Center for Economic Policy Research, Stanford University, January 1985; and Ken-ichi Imai, "Patterns of Innovation and Entrepreneurship in Japan" (Unpublished paper, Institute of Business Research, Hitotsubashi University, Tokyo, March 1988).

29. This discussion draws on interviews with Nobuyoshi Miyazawa of Enicom, Nippon Steel's software subsidiary, and Yasushi Suzuki of Nippon Steel Co., conducted in Tokyo by Martin Kenney, March 1989.

30. Aoki, *Information, Incentive and Bargaining in the Japanese Economy.*

31. Toyo Keizai Shinposha, *Japan Company Handbook* (Tokyo: Toyo Keizai Shinposha, 1988), 697.

32. On the product development orientation of Japanese R&D, see Henry Ergas, "Does Technology Policy Matter?" in Harvey Brooks and Bruce Guile, eds., *Technology and Global Industry* (Washington, D.C.: National Academy Press, 1987), 191–245; and Christopher Freeman, *Technology Policy and Economic Performance: Lessons from Japan* (London: Pinter Publishers, 1987). For a recent overview of Japanese R&D, see Gary Saxonhouse, "Technological Progress and R&D Systems in Japan and in the

NOTES

United States," in Uyehara, ed., *U.S.-Japan Science and Technology Exchange,* 29–56.

33. As reported in "Research and Development in Japan: 1989 Update," *JEI Reports,* no. 24A (June 23, 1989).

34. See Daniel Okimoto and Gary Saxonhouse, "Technology and the Future of the Economy," in Kozo Yamamura and Yasukichi Yasuba, eds., *The Political Economy of Japan* (Stanford, Calif.: Stanford University Press, 1988), 1:412.

35. "Assessing Japan's Role in Telecommunications," *IEEE Spectrum* (June 1986):48.

36. Quoted in Yasuzo Nakagawa, *Nihon no Handotai Kaihatsu* (The development of semiconductors in Japan) (Tokyo: Daiyamondo-sha, 1981), 4–5.

37. See "How Sony Pulled Off Its Spectacular Computer Coup," *Business Week,* January 15, 1990, 76–77.

38. See Masanori Moritani, *Japanese Technology* (Tokyo: Simul, 1982); and Terutomo Ozawa, *Japan's Technological Challenge to the West, 1950–1974* (Cambridge: MIT Press, 1974).

39. Quoted in U.S. Congress, House Subcommittee on Investigations and Oversight, and House Subcommittee on Science, Research, and Technology, *Japanese Technological Advances and Possible United States Responses Using Research Joint Ventures,* Hearings, 98th Cong., 1st Sess., June 29–30, 1983, 10. For a conceptual elaboration of this point, see Wesley Cohen and Daniel Levinthal, "Absorptive Capacity: A New Perspective on Learning and Innovation," *Administrative Science Quarterly* 34, 1 (March 1990):128–52.

40. Koji Kobayashi, *Computers and Communications: A Vision of C&C* (Cambridge: MIT Press, 1986); and Masaru Ibuka, "How Sony Developed Electronics for the World Market," *IEEE Transactions on Engineering Management* EM-22 (February 1975):15–19. The same basic pattern occurred in industry after industry. For automobiles, see David Halberstam, *The Reckoning* (New York: William Morrow, 1986). On steel, see Leonard Lynn, *How Japan Innovates: A Comparison with the U.S. in the Case of Oxygen Steel Making* (Boulder, Colo.: Westview Press, 1982); and Seiichiro Yonekura, "Recognizing Potential in Innovations: The Case of the U.S. and Japanese Steel Industries" (Discussion Paper no. 131, Institute of Business Research, Hitotsubashi University, Tokyo, July 1988).

41. There is a growing literature on this topic. See Hirotaka Takeuchi and Ikujiro Nonaka, "The New New Product Development Game," *Harvard Business Review* 64 (January–February 1986):137–46; Ken-ichi Imai, Ikujiro Nonaka, and Hirotaka Takeuchi, "Managing the New Product Development Process: How Japanese Companies Learn and

Unlearn," in Kim Clark, Robert Hayes, and Christopher Lorenz, eds., *The Uneasy Alliance: Managing the Productivity-Technology Dilemma* (Boston: Harvard Business School Press, 1985); Masahiko Aoki and Nathan Rosenberg, "The Japanese Firm as an Innovating Institution" (Working Paper no. 106, Center for Economic Policy Research, Stanford University, Stanford, Calif., 1987).

42. Takeuchi and Nonaka, "The New New Product Development Game," 137–38.

43. Nonaka highlights the importance of "redundancy" as a way of creating new ideas and products. In other words, the old conception of efficiency based on functional specialization and specific job descriptions is replaced by an overlapping team effort. Personal communications with Martin Kenney, December 1989.

44. This study tracked the careers of engineering graduates from Japan's Tohoku University and compared them with graduates of Carnegie Mellon University. See Leonard Lynn, Henry Piehler, and Walter Zaharay, *Engineers in the U.S. and Japan: A Comparison of Their Numbers and an Empirical Study of Their Careers and Methods of Information Transfer* (Pittsburgh, Pa.: Carnegie Mellon University, 1988).

45. See, for example, D. Eleanor Westney and Kiyonori Sakakibara, "Comparative Study of the Training, Careers and Organization of Engineers in Japan and the United States" (Unpublished working paper, MIT-Japan Science and Technology Program, Cambridge, Mass., 1985).

46. Dormitory life also generates close personal relationships that will become another conduit for information sharing as these young employees are transferred throughout the firm during their careers.

47. For a discussion of this, see Ikujiro Nonaka and Martin Kenney, "Innovation as an Organizational Information Creation Process: A Comparison of Canon Inc. and Apple Computer," *Hitotsubashi Business Review* (forthcoming, in Japanese). The not-invented-here syndrome is also weakened by rotation. Since an employee in one area may be moved to another, far less loyalty to a specific subsection of the company is built.

48. "Plus Development Corporation," Harvard Business School Case no. 9-687-001, 1986, 8–9.

49. See Kagono, Nonaka, Sakakibara, and Okumura, *Strategic versus Evolutionary Management,* 122.

50. See Dohse et al., "From 'Fordism' to 'Toyotism'?"; and Parker and Slaughter, "Management by Stress."

51. See Haruo Shimada and John Paul MacDuffie, "Industrial Relations and 'Humanware,' " (Working paper no. 1855-86, MIT, Sloan School of Management, Cambridge, Mass., December 1986). Kazuo Koike, "Skill Formation Systems in the U.S. and Japan: A Comparative Study," in Masahiko Aoki, ed., *Anatomy of the Japanese Firm* (Amsterdam: North-

Holland, 1984), 47–75; idem, "Human Resource Development and Labor-Management Relations," in Yamamura and Yasuba, eds., *Political Economy of Japan* 1:289–330; Kenneth Arrow, "The Economic Implications of Learning-by-Doing," *Review of Economic Studies* (1962).

52. Quoted in Myron Tribus, "Applying Quality Management Principles," *Research Management* 30 (November–December 1987):13

53. James Koford, LSI Logic, interview, April 1988.

54. See Ramchandran Jaikumar, "Postindustrial Manufacturing," *Harvard Business Review* 64 (November–December 1986):69–76; and Robert Hayes and Ramchandran Jaikumar, "Manufacturing's Crisis: New Technologies, Obsolete Organizations," *Harvard Business Review* 66 (September–October 1988):77–85.

55. This discussion is based on personal interviews and site visits we have conducted at Japanese automobile, steel, computer, and other transplant manufacturing facilities over the past three years. See Martin Kenney and Richard Florida, "Reindustrialization within Deindustrialization: Japanese Steel, Rubber and Automobile Production in the United States" (SUPA Discussion Paper no. 89-58, Carnegie Mellon University, Pittsburgh, Pa., November 1989); Andrew Mair, Richard Florida, and Martin Kenney, "The New Geography of Automobile Production: Japanese Transplants in North America," *Economic Geography* 64 (October 1988):352–73; Richard Florida and Martin Kenney, "Work Organization and Supplier Relations in the Japanese Transplants" (SUPA Discussion Paper no. 90–9, Carnegie Mellon University, Pittsburgh, Pa., January 1990).

56. On self-managing work teams, see Shimada and MacDuffie "Industrial Relations and 'Humanware.' " On Kyocera, see Gene Bylinsky, "The Hottest High Tech Company in Japan," *Fortune,* January 1, 1990, 83–88.

57. The physical layout of assembly lines helps this process. In Japan modular lines are used in place of the long dedicated transfer lines found in the United States. These arrangements include the use of more general-purpose machines that can be used for a variety of different production processes. These arrangements facilitated rapid shifts between different products within a product family. With this setup, lines can easily be converted to different products, and workers can perform a number of tasks on different machines simultaneously.

58. Site visit by authors and interview with Toshikata Amino, executive vice president of Honda America Manufacturing Corporation, April 1988.

59. See Shimada and MacDuffie, "Industrial Relations and 'Humanware.' " This point was repeatedly reinforced in our interviews with Japanese managers.

60. See Kagono et al., *Strategic versus Evolutionary Management.*

61. For a discussion of these momentous struggles, see Andrew Gordon, *The Evolution of Labor Relations in Japan: Heavy Industry, 1853–1955* (Cambridge: Harvard University Press, 1985); and Joe Moore, *Japanese Workers and the Struggle for Power, 1945–1947* (Madison: University of Wisconsin Press, 1983). For the implications of this for understanding the global political economy, see Martin Kenney and Richard Florida, "Beyond Mass Production: The Social Organization of Production and the Labor Process in Japan," *Politics and Society* 14, 2 (1988):121–58.

62. On the Japanese accord, see Kenney and Florida, "Beyond Mass Production," 121–58; Gordon, *Evolution of Labor Relations in Japan;* Moore, *Japanese Workers and the Struggle for Power.*

63. See Haruo Shimada, "The Perception and Reality of Japanese Industrial Relations," in Lester Thurow, ed., *The Management Challenge* (Cambridge: MIT Press, 1986), 42–68; and Taishiro Shirai, ed., *Contemporary Industrial Relations in Japan* (Madison: University of Wisconsin Press, 1983).

64. See Moore, *Japanese Workers and the Struggle for Power.*

65. In a recent paper Christoph Deutschmann argues that "the status of the individual worker in the firm does not primarily depend upon his work but on his age and length of service in the company. Leaving the firm would not only mean heavy financial losses . . . but also a loss of social status. In practice, therefore, the regular employee in a large company does not have 'exit' options and thus does not have an alternative to cooperating and coping with the company he has entered. It is this particular *limit* of security, not the security provided by lifetime employment itself (which in fact exists in other countries too), which is characteristic of the Japanese system" (emphasis in original). Christoph Deutschmann, "Economic Restructuring and Company Unionism: The Japanese Model," *Economic and Industrial Democracy* (November 1987):479.

66. See, for example, Yoko Kawashima and Toshiaki Tachibanaki, "The Effect of Discrimination and of Industry Segmentation on Japanese Wage Differentials in Relation to Education," *International Journal of Industrial Organization* 4 (1986):43–68.

67. Our position on these issues is more fully developed in a debate with Professors Tetsuro Kato and Rob Steven published in the Japanese journal *Mado* (Winter 1989, in Japanese).

68. James Abegglen and George Stalk, *Kaisha: The Japanese Corporation* (New York: Basic Books, 1985), 147.

69. Quoted in Janet Novack, "First You Borrow, Then You Innovate," *Forbes,* May 16, 1988, 73.

70. U.S. Department of Commerce, Patent and Trademark Office, "Tech-

nology Assessment and Forecast Database" (Unpublished data, January 1990).

71. "A Foreign Push in U.S. Patents," *New York Times,* June 4, 1989.

72. Computer Horizons, *Identifying Areas of Leading Edge Japanese Science and Technology* (Final Report to the Division of Science Resources Studies, National Science Foundation, Washington, D.C., April 15, 1988), 5.

73. National Science Foundation, *International Science and Technology Data Update: 1988* (Washington, D.C.: National Science Foundation, 1989), NSF 89-307, p. 7; Paula Doe, "Japan Spends Aggressively and Builds for the Future," *Electronic Business,* April 16, 1990, 77.

74. Interviews were conducted by Martin Kenney in June 1984, December 1987, June 1988, and October 1988 through March 1989. The responses were unanimously uniform in that R&D was considered the key to corporate survival.

75. See Kiyonori Sakakibara, "Global Technology Strategy of Japanese Firms" (Unpublished paper, Institute of Business Research, Hitotsubashi University, Tokyo, October 1988).

76. Many American commentators have called attention to Japan's technopolis program, which they see as an attempt by Japan to create its own versions of Silicon Valley. But according to leading Japanese industrialists and policymakers, the technopolis program is a regional economic development program designed to help revitalize declining regions and industries, that is, only a minor element in Japan's growing portfolio of technology development strategies. For a more traditional view, see Sheridan Tatsuno, *The Technopolis Strategy: Japan, High Technology, and the Control of the Twenty-first Century* (New York: Prentice-Hall, 1986).

77. Thus far, Hitachi, NEC, Kobe Steel, and Otsuka Pharmaceuticals have built new laboratories at U.S. universities. See "Advanced Bio Class: That's Over in Hitachi Hall," *Business Week,* August 7 1989, 73–74.

78. See Sheridan Tatsuno, *Created in Japan: From Imitators to World Class Innovators* (Cambridge, Mass.: Ballinger, 1990).

79. "Japan Keeps Up the Big Spending to Maintain Industrial Might," *New York Times,* April 11, 1990, 1.

80. See Kenney and Florida, "Beyond Mass Production"; and idem, "Japan's Role in a Post-Fordist Age," *Futures* (April 1988):136–51.

Chapter 9

1. As quoted in Frank Burger, "Commerce without Morality," *Electronic Business* (March 19, 1990):159.

2. See Rosabeth Kanter, *When Giants Learn to Dance* (New York: Simon & Schuster, 1989); and Tom Peters, *Thriving on Chaos: Handbook for a*

Management Revolution (New York: Knopf, 1987), for optimistic outlines of these efforts.

3. See Thomas Kochan, "Adaptability of the U.S. Industrial Relations System," *Science* 240 (April 15, 1988):287–92.
4. See "The Rival Japan Respects," *Business Week*, November 13, 1989, 108–18; "Motorola Sends Its Workforce Back to School," *Business Week*, June 6, 1988, 80–81; and Gary Stix, "At Motorola: New Alliances, New Management," *IEEE Spectrum* (July 1988):39–43.
5. On Xerox, see Thomas Kochan and Joel Cutcher-Gershenfeld, *Institutionalizing and Diffusing Innovations in Industrial Relations* (Report to the U.S. Department of Labor, Bureau of Labor-Management Relations and Cooperative Programs, Washington, D.C., 1988).
6. "The Payoff from Teamwork," *Business Week*, July 10, 1989, 56–62.
7. See "U.S. Industry's Unfinished Struggle," *New York Times*, February 21, 1988.
8. Kochan, "Adaptability of the U.S. Industrial Relations System," 289. See also Harry Katz, Thomas Kochan, and Jeffrey Keefe, "Industrial Relations and Productivity in the U.S. Automobile Industry," *Brookings Papers on Economic Activity* 3 (1987):685–727. For an excellent discussion of the necessity of providing workers real input in the organization of work, see Shoshana Zuboff, *In the Age of the New Machine* (New York: Basic Books, 1988).
9. Cited in Kochan, "Adaptability of the U.S. Industrial Relations System," 289.
10. "The Payoff from Teamwork," 58.
11. Harry Katz, "Business and Labor Relations Strategies in the U.S. Automobile Industry: The Case of General Motors" (Paper presented at the Conference on the Future of Work in the Automobile Industry, Berlin, West Germany, November 6, 1987).
12. "Payoff from Teamwork," 56.
13. For a critique of the "factory of the future" concept, see Bryn Jones, "When Certainty Fails: Inside the Factory of the Future," in Stephen Wood, ed., *The Transformation of Work* (London: Unwin Hyman, 1989), 44–58.
14. Cited in "High Tech to the Rescue: More Than Ever Industry Is Pinning Its Hopes on Factory Automation," *Business Week*, June 16, 1986, 100–108.
15. See "The Factory of the Future," *The Economist*, May 30, 1987, 5–18.
16. Ramchandran Jaikumar, "Postindustrial Manufacturing," *Harvard Business Review* 64 (November–December 1986):69–76; see also Robert Hayes and Ramchandran Jaikumar, "Manufacturing's Crisis: New Technologies, Obsolete Organizations," *Harvard Business Review* 66 (September–October 1988):77–85.

17. See Harley Shaiken, Stephen Herzenberg, and Sarah Kuhn, "The Work Process under More Flexible Production," *Industrial Relations* 52 (Spring 1986):167–83.

18. Alberto Socolovsky, "IBM Charlotte: Winner of the 1988 Factory Automation Award," *Electronic Business* (February 15, 1988):98–99.

19. Norm Alster, "Technology Transfer at General Motors," *Electronic Business* (June 1, 1988):33–40.

20. "Smart Factories: America's Turn," *Business Week,* May 8, 1989, 142–48.

21. Rick Wartzman and Andy Pasztor, "Stealth Bomber Comes under Fire in Congress as Its Price Mounts," *Wall Street Journal,* July 13, 1989, A1, A3.

22. See Stephen Roach, "America's Technology Dilemma: A Profile of the Information Economy," *Special Economic Study* (New York: Morgan Stanley, April 22, 1987); and idem, "Pitfalls on the 'New' Assembly-line: Can Services Learn from Manufacturing," *Special Economic Study* (New York: Morgan Stanley, June 22, 1989).

23. For more on the organizational dimensions of factory automation, see Maryellen Kelley and Harvey Brooks, *The State of Computerized Automation* (Cambridge, Mass.: John F. Kennedy School of Government, 1988).

24. Both examples are from Dwight Davis, "Beating the Clock," *Electronic Business* (May 29, 1989):21–28.

25. See, for example, Albert Link and Gregory Tassey, *Strategies for Technology-based Competition* (Lexington, Mass.: Lexington Books, 1987), 56–58.

26. See "An Open Door the U.S. Isn't Using," *Business Week,* May 15, 1989, 59–62; and Charles Owens, "Tapping Japanese Science," *Issues in Science and Technology* (Summer 1989):32–34.

27. See "A Corporate Lag in Research Funds Is Causing Worry," *New York Times,* January 23, 1990, 1, C6.

28. On Goodyear, see "High Price of Victory," *Financial Times,* December 15, 1988, 3.

29. See "As LBOs Take a Rocket Ride, R&D Takes a Nosedive," *Business Week,* February 27, 1989.

30. As quoted in *New York Times,* January 23, 1990, 1.

31. "Without Wall Street: High Tech Is Striking Deals, Foreign and Domestic," *Business Week,* 1989 special issue, "Innovation in America," 160.

32. See Mel Horwitch, "The Emergence of Value Creation Networks in Corporate Strategy," in Cecil Uyehara, ed., *U.S.-Japan Science and Technology Exchange: Patterns of Interdependence* (Boulder, Colo.: Westview Press, 1988), 188–215; and SRI International, "Strategic Partnerships: A New Corporate Response," (Report no. 730, SRI International, Business Intelligence Program, Stanford, Calif., 1986). For a less optimistic

view, see David Teece, "Profiting from Technological Innovation: Implications for Integration, Collaboration, Licensing and Public Policy," in Teece, ed., *The Competitive Challenge: Strategies for Industrial Innovation and Renewal* (Cambridge, Mass.: Ballinger, 1987), 201.

33. Harvey Brooks, statement before the U.S. Congress, House Committee on Science and Technology, *Japanese Technological Advances and Possible United States Responses Using Research Joint Ventures* (Washington, D.C., June 29–30, 1983), 19.

34. Eugene Kleiner, interview by authors, April 1988.

35. George Grodahl, partner in Broadview Associates, a San Francisco investment bank, quoted in Ray Wise, "The Electronics Cookie Jar is Loaded with M&A Goodies," *Electronic Business* (March 20, 1989):78.

36. See Martin Kenney, *Biotechnology: The University-Industrial Complex* (New Haven: Yale University Press, 1986), 163, for one such case.

37. As quoted in Edward Welles, "The Tokyo Connection," *Inc.* (February 1990):60. Dennis made much the same comments in an interview with us.

38. Our discussion of GE's takeover of Intersil is based on personal interviews with former GE and Intersil employees. See also Pamela Hamilton, "GE Launches Major LSI Technology Push," *Electronics* (February 24, 1981):95–99.

39. Name withheld by request, interview by authors, April 1988.

40. For the sale of the semiconductor group to Harris, see Janet Guyon, "GE Will Sell Its Chip Line to Harris Corp.," *Wall Street Journal,* August 16, 1988, 4; Alden Hayashi, "GE Says Solid State Is Here to Stay," *Electronic Business* (April 1988):53–56.

41. See Tom Quinlan, "Management Leadership Is Key to Zilog's Turnaround," *Electronic Business* (February 5, 1990):32–33.

42. It is interesting that Steven Jobs has until recently had nothing but contempt, sometimes verging on the pathological, for Japan and Japanese technology. See, for example, Frank Rose, *West of Eden* (New York: Viking, 1989).

43. See, Mark Dibner, "Strategic Alliances Are Likely to Remain Important in Japan's Biotech Industry," *Venture Japan* 2 (October 1989):47–52.

44. See "Without Wall Street: High Tech Is Striking Deals, Foreign and Domestic," 156–64.

45. Venture Economics (Published data, 1990).

46. John McCreadie and Philip Liv, "Wyse Deal May Signal a Boom in U.S. Taiwan Deals," *Electronic Business* (February 19, 1990):79–80.

47. See "Why Genentech Ditched the Dream of Independence," *Business Week,* February 19, 1990, 36–37.

48. "Stopping the High Tech Giveaway," *New York Times,* November 22, 1987, F1.

49. "Is the U.S. Selling Its High-Tech Soul to Japan," *Business Week,* June 26, 1989, 117.

50. See "Venture Capital Loses Its Vigor," *New York Times,* October 8, 1989.

51. As quoted in Ray Wise, "Wall Street Forces High Tech to Scramble for Capital," *Electronic Business* (May 29, 1989):37.

52. Yuichi Murano, Product Marketing Manager, Nihon Sun Microsystems, K.K., interview by Martin Kenney, March 1989. See Valerie Rice, "Breaking into Japan: Small U.S. Companies Find Success in a Demanding Market," *Electronic Business* (November 27, 1989):60–62; and "Small Companies Going Global," *New York Times,* November 27, 1989, C1.

53. On the potential rise of an offshore market for equity in U.S. start-up firms, see Wise, "Wall Street Forces High Tech to Scramble for Capital," 37–42.

54. Peter Brooke, interview by the authors, June 1988.

55. As reported in "Globalization Is the Driving Force," *Electronic Business* (March 19, 1990):44–48.

56. James Solomon, interview by the authors, April 1988; Tazz Pettibone quoted in Edward Welles, "The Tokyo Connection," *Inc.,* February 1990, 53.

57. Claudia Bird Schoonhoven and Kathleen Eisenhardt, *A Study of the Influence of Organizational, Entrepreneurial and Environmental Factors on the Growth and Development of Technology Based Startup Firms* (Report to the Economic Development Administration, Washington, D.C., 1987), 131.

58. "Is the U.S. Selling Its High-Tech Soul to Japan?" 117. Valentine echoed these concerns in an interview with us.

59. See Battelle Memorial Institute, *Research by Cooperative Organizations: A Survey of Scientific Research by Trade Associations, Professional and Technical Societies and Other Cooperative Groups* (Report to the National Science Foundation, Washington, D.C., 1956).

60. See Robert Noyce, "Cooperation Is the Best Way to Beat Japan," *New York Times,* July 9, 1989, Section 3, F2.

61. "How to Expand the Corporation," *Business Week,* April 11, 1983, 21.

62. MCC had ten charter members, twenty eventually joined, and three dropped out later. See Clifford Barney, "R&D Co-op Gets Set to Open Shop," *Electronics* (March 24, 1983); "MCC: An Industry Response to the Japanese Challenge," *IEEE Spectrum* (November 1983):55–56; Mark Fischetti, "A Review of Progress at MCC," *IEEE Spectrum* (March 1986):76–83; "MCC and Other Research Clubs: Do They Work?" *Electronic Business* (December 10, 1988):17–18; Morton Peck, "Joint R&D: The Case of the Microelectronics and Computer Technology Corporation," *Research Policy* 12 (1986):219–31.

63. See Valerie Rice and Carol Suby, "Sematech: United We Stand?" *Electronic Business* (May 1, 1988):30–37; Peter Waldman, "Sematech Rushes to Meet Japan Challenge," *Wall Street Journal,* January 8, 1988, 6; Mark Reagan and Michael Boss, "Sematech Goes to Austin—What's Next," *Dataquest Research Newsletter* (1988): No. 1988–2.

64. As quoted in Waldman, "Sematech Rushes to Meet Japan Challenge," 6.

65. "MCC and Other Research Clubs: Do They Work?" 17.

66. "Sematech Names Intel's Noyce to Head Semiconductor Industry Research Group," *Wall Street Journal,* July 28, 1988, 29.

67. Gordon Moore of Intel, as quoted in "This Will Surely Come Back to Haunt Us," *Business Week,* January 29, 1990, 72–73.

68. Rice and Suby, "Sematech: United We Stand?" 33.

69. Ibid., 35; Jack Robertson, "Tie Offshore Buys in U.S. to Trade Levels," *Electronic News,* May 14, 1990, 8.

70. "Sematech Today: Cash Dispenser," *New York Times,* January 4, 1990, C1.

71. For a discussion, see Richard Nelson, *High-Technology Programs: A Five Nation Comparison* (Washington, D.C.: American Enterprise Institute, 1984).

72. For a comparison of U.S., Japanese, and European R&D consortia, see John Alic, "Cooperation in R&D: When Does It Work?" (Paper presented to the Colloquium on International Cooperation between Rival Trading Nations, San Miniato, Italy, May 29–31, 1986).

73. For a history of university-industry relations, see David Noble, *America by Design: Science, Technology and the Rise of Corporate Capitalism* (New York: Oxford University Press, 1977).

74. See Martin Kenney, *Biotechnology: The University-Industrial Complex* (New Haven: Yale University Press, 1986).

75. "Where a Venture Capitalist Is Big Man on Campus," *Business Week,* July 28, 1986, 49.

76. See "A Biotech Venture Gives BU a Black Eye," *Business Week,* January 15, 1990, 32.

77. See Kenney, *Biotechnology: The University-Industrial Complex.*

78. Quoted in "More Partnerships to Boost Technology Sought for Academe," *The Chronicle of Higher Education* (April 19, 1989):A21–A22.

79. For further discussion, see Mark Crawford, "Utah Looks to Congress for Cold Fusion Cash," *Science* 244 (May 5, 1989):522–23.

80. As reported in Martin Tolchin "A Debate Over Access to American Research," the *New York Times,* December 17, 1989, F4.

81. See National Science Foundation, *University-Industry Relationships: Myths, Realities and Potentials* (Washington, D.C.: National Science Foundation, 1982). On the Engineering Research Center, see U.S. General Accounting Office, *Engineering Research Centers: NSF Program Management and*

Industry Sponsorship (Washington, D.C.: General Accounting Office, 1988).

82. See, for example, Philip Shapira, "Modern Times: Learning from New State Initiatives in Industrial Extension and Manufacturing Technology Transfer" (Paper presented at the Annual Conference of the Association of Collegiate Schools of Planning, Buffalo, New York, October 1988).

83. As cited in John Ullman, "Economic Conversion: Indispensable for America's Economic Recovery," (National Commission for Economic Reconversion and Disarmament, Washington, D.C., 1989).

84. Judith Larsen, "Technology Transfer: Can America Learn to Move Knowledge," *Dataquest Strategic Issue* (June 1988).

85. Harvey Brooks and Lewis Branscomb, "Rethinking the Military's Role in the Economy," *Technology Review* (August–September 1989):56.

86. Lynn Browne, "Defense Spending and High-Technology Development: National and States Issues," *New England Economic Review* (September–October 1988):3–22.

87. On the concept of "technological trajectories," see Giovanni Dosi, "Technological Paradigms and Technological Trajectories," *Research Policy* 2 (June 1982):147–62; and Christopher Freeman and Carlotta Perez, "Structural Crises of Adjustment," in Giovanni Dosi et al., eds., *Technical Change and Economic Theory* (London: Pinter Publishers, 1988), 38–66.

88. See Herb Brody, "Star Wars: Where the Money's Going," *High Technology Business* (December 1987):22–29.

89. See Frank Lichtenberg, "Military R&D Depletes Economic Might," *Wall Street Journal,* August 26, 1986.

90. See Ullman, "Economic Reconversion: Indispensable for America's Economic Recovery."

91. Michael Leibowitz, "Does Military R&D Stimulate Commerce or the Pork Barrel?" *Electronic Business* (February 6, 1989):54.

92. For background on these new state programs, see Scott Fosler, *The New Economic Role of American States* (New York: Oxford University Press, 1988); Edward Morrison, "State and Local Efforts to Encourage Economic Growth through Innovation: An Historical Perspective," in Denis Gray, Trudy Solomon, and William Hetzner, eds., *Technological Innovation: Strategies for a New Partnership* (Amsterdam: North-Holland, 1986), 57–67; Walter Plosila, "State Technical Development Programs," *Forum for Applied Research and Public Policy* (Summer 1987):270–79; Charles Watkins and Joan Wills, "State Initiatives to Encourage Economic Development through Technological Innovation," in Gray, Solomon, and Hetzner, eds., *Technological Innovation: Strategies for a New Partnership.*

93. On state venture capital, see Richard Florida and Donald Smith, "Venture Capital, Innovation and Economic Development," *Economic Development Quarterly* (November, 1990, forthcoming); on science parks, see Michael Luger and Harvey Goldstein, "Science-Research Parks as In-

struments of Technology-Based Regional Policy: An Assessment" (Paper presented at the North American meetings of the Regional Science Association, Toronto, Canada, November 1988).

94. These data are from Minnesota Department of Trade and Economic Development, *State High Technology Programs in the United States* (Minneapolis, Minn.: 1988).

95. David Osborne, *Laboratories of Democracy: A New Breed of Governor Creates Models for National Growth* (Boston: Harvard Business School Press, 1988).

96. Professors hired by these consortia are also encouraged to switch research to more applied topics. Further, much of the money invested in these programs is often rerouted from less commercially oriented departments and professors.

97. See Bennett Harrison and Sandra Kanter, "The Political Economy of States' Job Creation Business Incentives," *Journal of the American Institute of Planners* (October 1978):424–35, for a general discussion of interstate, interjurisdictional competition

98. For an overview of intellectual property issues, see Judith Larsen and John Wilson, "Intellectual Property: The New Crown Jewels," *Dataquest Strategic Issue* (March 1988).

99. Eric Larson, "In High-Tech Industry, New Firms Often Get Fast Trip to Courtroom," *Wall Street Journal,* August 14, 1984, 1, 7.

100. James Miller, "Big Firms Pursue Start-ups with Suits," *Wall Street Journal,* February 24, 1989, A25.

101. Jeffrey S. Young, *Steve Jobs: The Journey Is the Reward* (Glenview, Ill.: Scott, Foresman, 1988).

102. Valerie Rice, "Power of Attorney: How Intel Defends Its Technology Rights," *Electronic Business* (September 1, 1988):31–38.

103. Alden Hayashi, "Is DEC Using Litigation as a Marketing Tool?" *Electronic Business* (September 1, 1988):46.

104. Nancy Brumback, "Preventive Law: Protecting Your Interests Beforehand," *Electronic Business* (May 15, 1988):74.

105. Richard Schmitt, "Apple Wins First Round in Software-Copyright Case," *Wall Street Journal,* March 20, 1989, B1.

106. On Lotus, see Lawrence Tell, "Software Copyrights: Keep Out the Pirates—But Let Innovators In," *Business Week,* August 31, 1987, 31. For a more general discussion, see Kathleen Wiegner and John Heins, "Can Las Vegas Sue Atlantic City?" *Forbes,* March 6, 1989, 130–37.

107. Miller, "Big Firms Pursue Start-ups with Suits," A25. See also Hayashi, "Is DEC Using Litigation as a Marketing Tool?"

108. See Alvin Klevorick, Richard Nelson, and Sidney Winter, "Appropriating the Returns from Industrial R&D," *Brookings Papers on Economic Activity* (1987):783–820.

109. Teece, "Profiting from Technological Innovation," 188–89.

110. Hitachi officials are certain they can prove other patent infringements by Motorola in pending law suits. See Robert Ristelhueber, "Motorola Wins Stay of Ruling Blocking Sales of 68030 MPU," *Electronic News* (April 2, 1990); "Motorola, Hitachi Reach a Draw in Patents-Rights Suits," *Wall Street Journal,* March 30, 1990; and, "Hitachi, Motorola Locked in Chip-Patent Dispute," *Wall Street Journal,* January 27, 1989.

111. Stuart Zipper, "Apple Survival Hangs on 68030 Decision," *Electronic News,* May 7, 1990, 1, 31.

112. For an illuminating discussion of this topic, see David Jeremy, *Transatlantic Industrial Revolution: The Diffusion of Textile Technologies between Britain and America, 1790–1830's* (Cambridge: MIT Press, 1981).

Chapter 10

1. As chapter 9 shows, there are a huge number of advocates for such a strategy. For recent examples of such proposals, see "U.S. Aid Sought for Electronics," *New York Times,* October 30, 1989; "White House Pushes for Advances in Supercomputing," *Electronic Business* (October 30, 1989). Academic rationales for such proposals can be found in Charles Ferguson, "From the People Who Brought You Voodoo Economics" *Harvard Business Review* 66 (May–June 1988); and Michael Borrus, *Competing for Control* (Cambridge, Mass.: Ballinger, 1988).

2. See, for example, the Business Roundtable, *American Excellence in a World Economy* (New York: Business Roundtable, 1987); the Cuomo Commission on Trade and Competitiveness, *The Cuomo Commission Report* (New York: Simon & Schuster, 1988); the President's Commission on Industrial Competitiveness, *Global Competition: The New Reality* (Washington, D.C.: U.S. Government Printing Office, 1985); Michael Dertouzos et al., *Made in America* (Cambridge: MIT Press, 1989). An excellent critique of the MIT report, which provides an overview of a series of alternative macroeconomic and institutional approaches to restructuring, is contained in Jonathan Schlefer, "Making Sense of the Productivity Debate: A Reflection on the MIT Report," *Technology Review* (August–September 1989):28–40.

3. See Robert Costello, *Bolstering Defense Industrial Competitiveness: Report to the Secretary of Defense by the Under Secretary of Defense (Acquisition)* (Washington, D.C.: Office of the Under Secretary of Defense for Acquisitions, Department of Defense, 1988).

4. See, for example, the MIT report as well as recent studies published by the Office of Technology Assessment and the General Accounting Office; Dertouzos et al., *Made in America;* Office of Technology Assessment, *The Defense Industrial Base: Introduction and Overview* (Washington, D.C.: 1988); U.S. General Accounting Office, *Industrial Base: Defense*

Critical Industries (Washington, D.C.: U.S. General Accounting Office, 1988). For a critical review (from a Japanese perspective), see Japan Economic Institute, "The Defense Base Initiative: Opening the Door for Technological Protection," *JEI Report* 24A (Washington, D.C.: Japan Economic Institute, June 24, 1988).

5. Dertouzos et al., *Made in America,* 41.

6. On this point, see Harvey Brooks and Lewis Branscomb, "Rethinking the Military's Role in the Economy," *Technology Review* (August–September 1989).

7. On the overwhelming defense orientation of domestic public policy, see Seymour Melman, *The Permanent War Economy* (New York: Touchstone, 1985); and Alan Wolfe, *America's Impasse: The Rise and Fall of the Politics of Growth* (New York: Pantheon, 1981).

8. See the cover story "The Peace Economy: How Defense Cuts Will Fuel America's Long-Term Prosperity," *Business Week,* December 11, 1989, 50–55.

9. U.S. Congress, House of Representatives, H.R. 101, *A Bill to Facilitate the Economic Adjustment of Communities, Industries and Workers to Reductions and Realignments in Defense or Aerospace Contracts, Military Facilities and Arms Export, and for Other Purposes,* 101st Cong., 1st Sess., January 3, 1989. See also Seymour Melman, *The Demilitarized Economy: Disarmament and Conversion* (Montreal: Harvest House, 1988).

10. See "Statement of Representative Ted Weiss before the House Banking Subcommittee on Economic Stabilization on Title II of the Defense Production Act Pertaining to Economic Conversion," June 13, 1989. For a more detailed discussion of the underlying rationale of such policy, see Jonathan Feldman, Robert Krinksy, and Seymour Melman, "Criteria for Economic Conversion Legislation" (Briefing Paper no. 4, National Commission for Economic Conversion and Disarmament, Washington, D.C., December 1988).

11. Michael Cooley, *Architect or Bee* (Boston: South End Press, 1980). For discussions of conversion strategies developed in other European countries, see Lloyd Dumas and Marek Thee, eds., *Making Peace Possible: The Promise of Economic Conversion* (Elmsford, N.Y.: Pergamon Press, 1989), which has chapters on Britain, Sweden, and Norway.

12. For two illustrative statements, see Shoshana Zuboff, *In the Age of the Smart Machine* (New York: Basic Books, 1988); and Ramchandran Jaikumar, "Postindustrial Manufacturing," *Harvard Business Review* 64 (November–December 1986):69–76.

13. See Tessa Morris-Suzuki, "Robots and Capitalism," *New Left Review* 147 (1984).

14. Art Fong, interview by authors, April 1988. Fong's comments were echoed by Alan Bagley, very early original Hewlett-Packard engineer.

Index

■

253

INDEX

INDEX

INDEX

INDEX